COMPUTATIONAL APPROACHES TO CONSCIOUS ARTIFICIAL INTELLIGENCE

Series on Machine Consciousness

ISSN: 2010-3158

Series Editor: Antonio Chella *(University of Palermo, Italy)*

Published

Vol. 5   *Computational Approaches to Conscious Artificial Intelligence*
edited by Antonio Chella

Vol. 4   *Consciousness and Robot Sentience (Second Edition)*
by Pentti O Haikonen

Vol. 3   *The Revolutions of Scientific Structure*
by Colin G Hales

Vol. 2   *Consciousness and Robot Sentience*
by Pentti O Haikonen

Vol. 1   *Aristotle's Laptop: The Discovery of our Informational Mind*
by Igor Aleksander and Helen Morton

Series on Machine Consciousness – Vol. 5

COMPUTATIONAL APPROACHES TO CONSCIOUS ARTIFICIAL INTELLIGENCE

Editor

Antonio Chella

University of Palermo, Italy & ICAR-CNR Palermo, Italy

World Scientific

NEW JERSEY · LONDON · SINGAPORE · BEIJING · SHANGHAI · HONG KONG · TAIPEI · CHENNAI · TOKYO

Published by

World Scientific Publishing Co. Pte. Ltd.

5 Toh Tuck Link, Singapore 596224

USA office: 27 Warren Street, Suite 401-402, Hackensack, NJ 07601

UK office: 57 Shelton Street, Covent Garden, London WC2H 9HE

British Library Cataloguing-in-Publication Data
A catalogue record for this book is available from the British Library.

Series on Machine Consciousness — Vol. 5
COMPUTATIONAL APPROACHES TO CONSCIOUS ARTIFICIAL INTELLIGENCE

ISBN 978-981-127-666-8 (hardcover)
ISBN 978-981-127-667-5 (ebook for institutions)
ISBN 978-981-127-668-2 (ebook for individuals)

For any available supplementary material, please visit
https://www.worldscientific.com/worldscibooks/10.1142/13421#t=suppl

Contents

Preface

Despite strong claims in the popular press and social networks, AI has not developed machine consciousness yet. AI scholars have, however, developed computational models that take their cues from aspects of natural consciousness. These models take inspiration from the fact that consciousness is related to an integrated experience; we perceive the natural world as a whole of sounds, colors, shapes, and images. Other models are based on working memory which is critical for the global perception of the external world. Furthermore, models of attention are receiving great impetus, as a capability of consciousness is to perceive the world as a whole and focus on something in detail. Finally, an exciting area of research is the analysis of physical models, such as quantum and free energy-based models. After all, consciousness is a phenomenon of the physical world, and we must be able to study it with the tools proper to physics.

Biology may also contribute to building machine consciousness in that biological models inspire the computational models underlying the operations of conscious machines. The fact that consciousness may be related to an integrated experience, attention, and working memory are all biological suggestions. Nobel laureate Gerald Edelman pioneered using robots to validate brain organization and biological consciousness theories. The guiding principle is that we only understand something well if we can construct it. So, we will understand natural consciousness well when we can build artificial consciousness.

The study of bioengineering and bioinformatics is greatly aiding the understanding of biological processes and neuroscience, which underlie consciousness. For example, today, researchers have access to much brain data from MRI, CT scans, electroencephalograms, etc. AI helps find computational models to explain these data and hopefully predict new data.

The debate about AI and consciousness today is alive and vital. An important topic concerns whether consciousness can be implemented on a computer or robot or is something that escapes computational models and belongs only to the world of biology. Then, only a living being can be conscious. Therefore a robot, no matter how sophisticated, never will be.

Another debate concerns the role of embodiment in consciousness: can a bodiless being such as a computer be conscious, or is it necessary for a body with sensors to see and actuators to be able to move, touch, look, and so on? According to this hypothesis, a robot could be conscious, while a computer can never be conscious. However, this hypothesis has not been proven, and many scholars have rejected it.

A related issue concerns the role of the external environment: does consciousness reside in the brain, or is it relative to the environment around us? These issues are essential because, brought back to the human being, they would deny the ability of consciousness to the brain only, without a body and an external environment.

Furthermore, the big question is: can a robot feel something one day, such as pain or joy, or will it only be able to fake it? A related debate concerns the measurement of consciousness: how will we know if a robot is genuinely conscious? It may not be conscious, yet it could flawlessly imitate the experience of feeling emotions. It is also true for human beings: how do we know that another person is conscious? The current methods largely revolve around variations of the Turing test, which is based on the theme of imitation. This test has been subjected to extensive criticism: it requires a third conscious inidividual to evaluate the robot's performance, it relies entirely on language, and yet a conscious robot might not possess language capabilities, among other issues. Moreover, the latest language generation systems, such as GPT-3, which are not conscious, could almost pass the test because of their richness of language.

The ethical issue is today a hot topic in Artificial Intelligence studies. The problem may be closely related to the problem of consciousness. If a natural or artificial entity cannot feel something, it can never distinguish right from wrong.

All the laws and rules humanity has given may be incomprehensible and appear to be mere rhetorical conventions for an entity incapable of suffering and rejoicing. Asimov's famous three laws of robotics are fully supportable, but like all laws, they hide many ambiguities, and in fact, Asimov's exciting stories take their cue precisely from these ambiguities. A conscious entity that feels no emotion cannot understand what is good and evil beyond laws and rules. Then, the ethical problem in AI, or the problem of aligning the values of the machine with those of the person, can only be solved if the problem of making the robot feel something is solved simultaneously. On the other side, building a genuinely conscious machine may mean creating an entity capable of suffering. Then, the designers of a conscious machine would be responsible for increasing the suffering in the world, as they would be subjecting even machines and robots to potential distress.

Another critical topic concerns building the mind not of an adult but of a child and then making it grow and evolve. It is a widespread challenge in robotics, dating back to Alan Turing. However, today's learning techniques are mathematical optimization procedures with little in common with natural learning ability. Moreover, besides being intelligent, the machine should be able to develop some form of consciousness.

The field of machine consciousness is an aggregator for international research collaborations. The U.S. and Asian research institutions are very active in this new strand of research. In addition to academic research, there are startups in Silicon Valley active on this topic. European research in universities and startups is also interested in this topic, with initiatives and dedicated project calls. So many initiatives have arisen recently in the United Kingdom as well.

The challenge of building a conscious entity is a formidable task that may probably be pursued not by a single researcher or a research lab but by a combined effort of many computer scientists, philosophers, cognitive scientists, and neuroscientists. This book presents some of the computational aspects of machine consciousness from recognized scholars on this research topic. All the chapters originate from extended and fully revised papers initially published in the Journal of Artificial Intelligence and Consciousness, the leading journal in the field.

Palermo, May 13, 2023   Antonio Chella

Chapter 1

The Conscious Machine

Yan H. Ng[1], Antonio Chella[2]

[1]World Scientific Publishing Company, Singapore
[2]Università degli Studi di Palermo & ICAR-CNR Palermo, Italy

Conscious is a Latin word whose original meaning was "knowing" or "aware." So, a conscious person has an awareness of her environment and her own existence and thoughts. If you're "self-conscious," you're overly aware and even embarrassed by how you think you look or act, then could a machine be "aware" or "knowing what it is about to do?"

A "machine" is any causal physical system; therefore, we are machines, hence machines can be conscious (feel). The question is: which "kinds" of machines can feel, and how? Chances are those robots that should have passed the Turing Test — completely indistinguishable from us in their behavioral capacities — can feel, but we can never be sure. And we can never explain or understand "how" or "why" they manage to feel, if they do, because of the "mind/body" problem. We can only know how they pass the Turing Test. That is why this problem is not just "hard" — it is insoluble.

Based on the five skandhas in Buddhism that have comprehensively summed up the human cognitive, perceptual process put forward 2,500 years ago; namely, form, sensations, perception, mental formation, and consciousness, we review the nine articles in this book covering computational approaches to conscious artificial intelligence to investigate where they fit in into these five-stages human cognitive process.

1) Form (or material image, impression) (rupa)
2) Sensations (or feelings, received from form) (vedana)
3) Perceptions (samjna)
4) Mental activity or formations (sankhara)
5) Consciousness (vijnana)

1

While Human Intelligence aims to adapt to survive in a new environment by utilizing a combination of different cognitive processes, Artificial Intelligence aims to build machines that can mimic human behavior and perform human-like actions. The human brain is analogous, but machines are digital. However, by studying the process of human cognition will surely enlighten us in the understanding of machine Intelligence to develop fully autonomous unmanned systems to be conscious to aware of the problem, the object in the environment to feel, to recognize the object and discrimination of its components and aspects that supports all experience corresponding phenomena associate with this object, mimic humans to develop solution to overcome the problem.

Keywords: form (rupa), sensations (vedana), perceptions (samjna), mental formations (sankhara), consciousness (vijnana)

1. Introduction

Consciousness, at its simplest, is sentience and awareness of internal and external existence. However, the lack of definitions has led to millennia of analyses, explanations, and debates by philosophers, theologians, linguists, and scientists. Opinions differ about what exactly needs to be studied or even considered consciousness.

The dictionary definitions of the word *consciousness* extend through several centuries. They reflect a range of seemingly related meanings, with some controversial differences, such as the distinction between 'inward awareness' and 'perception' of the physical world, the distinction between 'conscious' and 'unconscious', or the notion of a "mental entity" or "mental activity" that is not physical.

The common usage definition of *consciousness* in Cambridge Dictionary defines consciousness as *"the state of understanding and realizing something."*

The Oxford Living Dictionary defines consciousness as *"The state of being aware of and responsive to one's surroundings"*, *"A person's awareness or perception of something"* and *"The fact of awareness by the mind of itself and the world."*

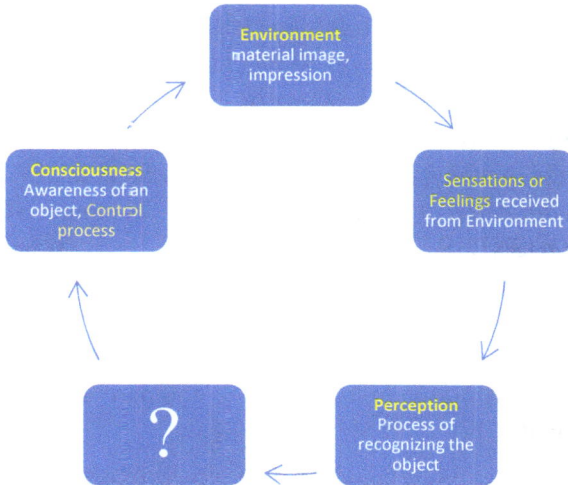

Fig. 1. The Definition of Consciousness from Oxbridge Dictionaries.

These two definitions seem to agree in one key aspect of consciousness, related to awareness of the environment surrounding the object that caught one's attention [Sensation + Memories + Thought Processes] with the awareness of an object in the environment through sight, sounds, touch, smell, and taste, that perhaps trigger one's memory or experiences with the object concerned, but did not extend to cover the consequence of this awareness.

In *Webster's Third New International Dictionary,* consciousness is defined as something that it touches but does not include the missing state of Mental Formations, which is often seen as the results of *emotion, volition, or thought; mind in the broadest possible sense trigger by the awareness or perception*:

1. *Awareness or perception of an inward psychological or spiritual fact; intuitively perceived knowledge of something in one's inner self*
   o *inward awareness of an external object, state, or fact*
   o *concerned awareness;* INTEREST, CONCERN — *often used with an attributive noun [e.g., class consciousness]*

2. *The state or activity that is characterized by sensation, emotion, volition, or thought; mind in the broadest possible sense; something in nature that is distinguished from the physical*
3. *The totality in psychology of sensations, perceptions, ideas, attitudes, and feelings of which an individual or a group is aware at any given time or within a particular time span* — compare STREAM OF CONSCIOUSNESS
4. *Waking life (as that to which one returns after sleep, trance, fever) wherein all one's mental powers have returned* . . .
5. *The part of mental life or psychic content in psychoanalysis that is immediately available to the ego* — compare PRECONSCIOUS, UNCONSCIOUS.

Nevertheless, philosophers have attempted to clarify technical distinctions using a jargon of their own. The *Routledge Encyclopedia of Philosophy* in 1998 defines consciousness as follows:

Consciousness — Philosophers have used the term 'consciousness' for four main topics:

- Knowledge in general, intentionality, introspection (and the knowledge it specifically generates) and phenomenal experience... Something within one's mind is 'introspectively conscious' just in case one introspects it (or is poised to do so). Introspection is often thought to deliver one's primary knowledge of one's mental life.
- An experience or other mental entity is 'phenomenally conscious' just in case there is 'something it is like' for one to have it.
- The clearest examples are: perceptual experience, such as tasting and seeing; bodily-sensational experiences, such as those of pains, tickles and itches; imaginative experiences, such as those of one's own actions or perceptions; and streams of thought, as in the experience of thinking 'in words' or 'in images'. Introspection and phenomenality seem independent, or dissociable, although this is controversial.

Many philosophers and scientists have been unhappy about the difficulty of producing a definition that does not involve circularity or fuzziness.

In The *Macmillan Dictionary of Psychology* (1989 edition), Stuart Sutherland expressed a skeptical attitude more than a definition:

Consciousness — The having of perceptions, thoughts, and feelings; awareness. The term is impossible to define except in terms that are unintelligible without a grasp of what consciousness means. Many fall into the trap of equating consciousness with self-consciousness — to be conscious it is only necessary to be aware of the external world. Consciousness is a fascinating but elusive phenomenon: it is impossible to specify what it is, what it does, or why it has evolved. Nothing worth reading has been written on it. However, this version of definition of consciousness fills the gap of what is missing in Fig. 1.

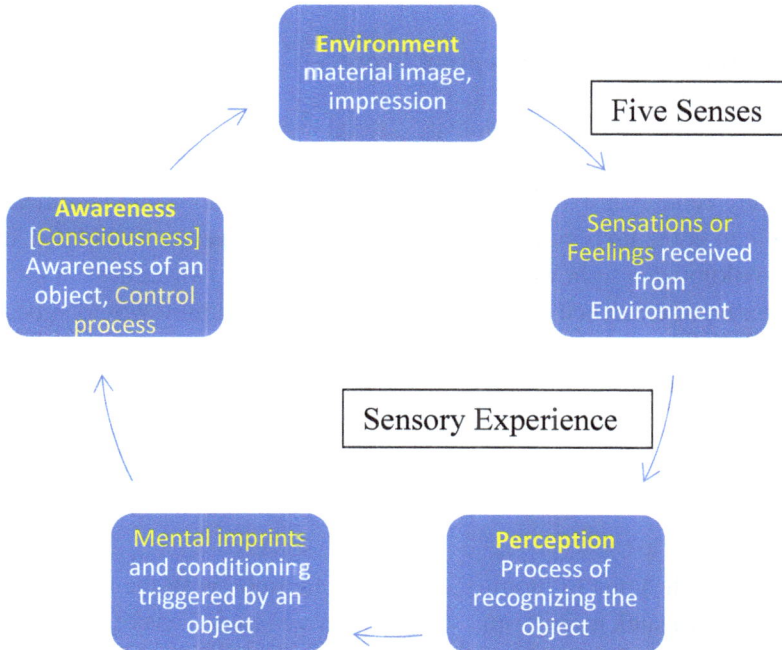

Fig. 2. Consciousness according to The *Routledge Encyclopedia of Philosophy*.

A partisan definition such as Sutherland's can hugely affect researchers' assumptions and the direction of their work: If awareness of the environment . . . is the criterion of consciousness, then even the protozoans are conscious. If awareness of awareness is required, then it is doubtful whether the great apes and human infants are conscious.

Many philosophers have argued that consciousness is a unitary concept that is understood by the majority of people despite the difficulty philosophers have had defining it. Others, though, have argued that the level of disagreement about the meaning of the word indicates that it either means different things to different people (for instance, the objective versus subjective aspects of consciousness), that it encompasses a variety of distinct meanings with no simple element in common, or that we should eliminate this concept from our understanding of the mind, a position known as consciousness semanticism.

2. Human Cognitive Perceptual Process

However, cognitive psychologists (Sternberg, 2012) take a more subtle view. Not everything we do, reason, and perceive is necessarily conscious. We may be unaware of stimuli that alter our perceptions and judgments or unable to come up with the right word in a sentence even though we know that we know the right word. However, sentience and awareness of internal and external existence is the bottom-line definition they would subscribe to, However, they believe consciousness includes both the feeling of awareness and the content of awareness. Therefore, attention and consciousness form two partially overlapping sets.

Conscious attention serves three purposes in playing a causal role in cognition.

- Firstly, it helps in monitoring our interactions with the environment. Through such monitoring, we maintain our awareness of how well we are adapting to the situation in which we find ourselves.
- Secondly, it assists us in linking our past (memories) and our present (sensations) to give us a sense of continuity of experience. Such continuity may even serve as the basis for personal identity.

- Thirdly, it helps us in controlling and planning for our future actions. We can do so based on the information from monitoring and from the links between past memories and present sensations.

At any point in time, we perceive a lot of sensory information. Through attentional processes (which can be automatic or controlled), we filter out the information that is relevant to us and that we want to attend to. Eventually, this leads to our taking action on the basis of the information we attended to.

In addition to the definitions stated in some of the dictionaries of consciousness related to awareness of the environment surrounding the object that caught one attention [Sensation + Memories + Thought Processes], cognitive psychologists include the consequence of consciousness that lead to [Control Processes + Automatic Processes].

This description views consciousness as a process consisting of five stages is a concept put forward 2,500 years ago as the Five skandhas in Buddhism that have comprehensively summed up this human cognitive, perceptual process. Since Cognitive Psychology is the study of how people perceive, learn, remember, and think about information; in order to learn and understand how we perceive? How we learn? How we remember and how we think, will, no doubt, provide us with the clue on the development tools and methodology for machines to mimic us to perform human-like actions.

1. Form (or material image, impression) (*rupa*)
2. Sensations (or feelings, received from form) (*vedana*)
3. Perceptions (*samina*)
4. Mental activity or formations (*sankhara*)
5. Consciousness (*vijnana*)

In Buddhism, the five skandhas are essentially a method for understanding that every aspect of our lives is a collection of constantly changing experiences. There is no one aspect that is truly solid, permanent, or unique. Everything is in flux. Everything is *dependent* upon multiple causes and conditions.

For example, *rupa-skandha* [Form] refers to everything in our material world — our body and our physical surroundings. All these things are constantly changing. *Vedana-skandha* refers to our sensations that are positive, negative, or indifferent–all our sensations are fleeting, changing from moment to moment. *Samjna-skandha* refers to all of our recognition or labeling of everything that we see, hear, smell, touch, or think; these labels are also constantly in flux. *Samskara-skandha* (all of our mental habits, thoughts, ideas, opinions, prejudices, compulsions, etc.) are dependent on many causes and conditions and always changing. Our consciousness itself is not one single thing but a collection of consciousness (*vijnana-skandha*) that is also constantly changing.

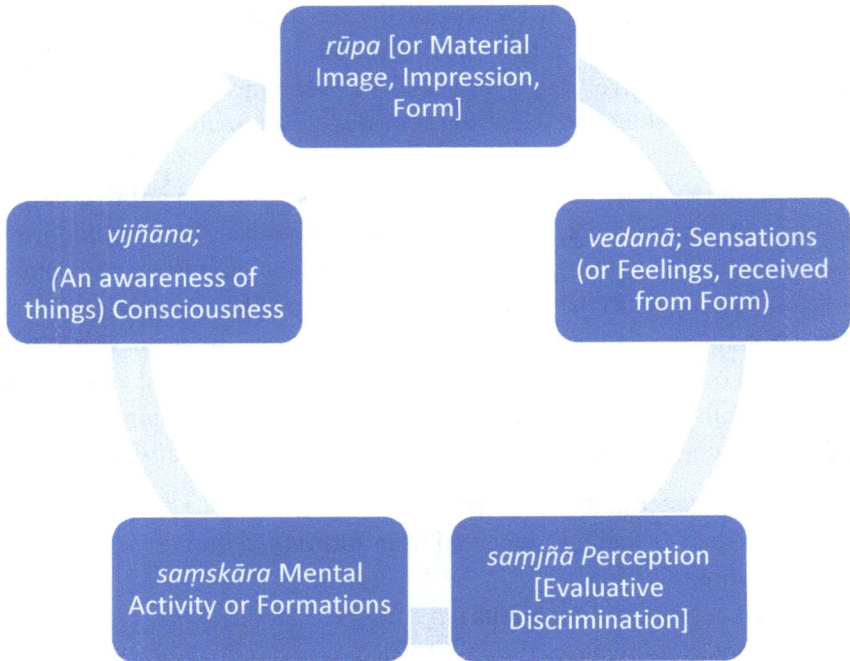

Fig. 3. The human cognitive perceptual process.

1. **"Form"** or **"Matter"** (*rūpa*): matter, body, or "material form" of a being or any existence. It includes our own bodies, and material objects as well. Specifically, the aggregate of form includes the five physical organs (eye, ear, nose, tongue, body) and our relationship

with our environment through the medium of our bodies and senses, (sight, sound, smell, taste, and tangible objects). In the context of machine learning, we often refer to Form as the environment. This is matter that is tangible (i.e. can be touched).

2. "**Sensation**" or "**Feeling**" (*vedanā*): the sensory experience of an object. In general, *vedanā* refers to the pleasant, unpleasant, and neutral sensations that occur when our internal sense organs come into contact with external sense objects and the associated consciousness. The sensations you experience in your body, including all pain and pleasure. These are feelings experienced from and through our five senses.

3. "**Perception**" (*saṃjñā*): the sensory and mental process that registers, recognizes, and labels (for instance, the shape of a tree, color green, emotion of fear). To receive information from the environment ["material form" of a being or any existence] using the sensory experience of an object to distinguish its features or characteristics. The process of recognising what things are? *Samjna* is the faculty that recognizes. Most of what we call thinking fits into the aggregate of samjna. The word "samjna" means "knowledge that puts together." It is the capacity to conceptualize and recognize things by associating them with other things. For example, we recognize shoes as shoes because we associate them with our previous experience with shoes. When we see something for the first time, we invariably flip through our mental index cards to find categories we can associate with the new object. It's a "some kind of tool with a red handle," for example, putting the new thing in the categories "tool" and "red." Or, we might associate an object with its context. We recognize an apparatus as an exercise machine because we see it at the gym.

4. "**Mental Formations**" (*saṃskāra*): "constructing activities", "conditioned things", "volition", "karmic activities"; all types of mental imprints and conditioning triggered by an object. Includes any process that makes a person initiate action or act. All volitional actions, good and bad, are included in the aggregate of mental formations, or *samskara*. How are actions "mental" formations? Thoughts.

5. **"Consciousness"** (*vijñāna*): *"discrimination"* or *"discernment"*. Awareness of an object and discrimination of its components and aspects that support all experience and corresponding phenomena as its object.

    It is important to understand that this awareness or consciousness depends on the other skandhas and does not exist independently from them. It is awareness but not recognition, as recognition is a function of the third skandha. This awareness is not a sensation, which is the second skandha. An awareness of things.

3. The five Skandhas as the Model for Human Cognition

(1) **Form:** or the environment in this case (or material image, impression)

All things that our five senses could perceive, or information that could be captured for the purposes of learning and recognition. Hence, the very early days of research and development in artificial intelligence to mimic human behavior and perform human-like actions of seeing started in the form of signal and image processing as the pretext for perception: How do we make sense of what we see?

The field of signal and image processing encompasses the theory and practice of algorithms and hardware that convert signals produced by artificial or natural means into a form useful for a specific purpose. The signals might be speech, audio, images, video, sensor data, telemetry, electrocardiograms, or seismic data, among others; possible purposes include transmission, display, storage, interpretation, classification, segmentation, or diagnosis. Researchers in this field span the areas of digital signal processing, statistical signal processing, image/video compression, analysis & processing, speech processing, music information retrieval, and computer audition.

Current research in digital signal processing includes robust and low-complexity filter design, signal reconstruction, filter bank theory, and wavelets. In statistical signal processing, researchers' interests include adaptive filtering, learning algorithms for neural networks, spectrum estimation and modeling, and sensor array processing with applications in sonar and radar. Image processing work is in restoration, compression, quality evaluation, computer vision, and medical imaging. Speech processing research includes modeling, compression, and recognition. Hence, with the help of signal and image processing, the data from the environment [form] are collected and captured for the purpose of recognition.

(2) **Sensations:** (or feelings received from form)

An act of recognizing or the state of being recognized. The identification of something as having been previously seen, heard, known, etc. The perception of something as existing or true; realization. The acknowledgment of something as valid or as entitled to consideration: the recognition of a claim. Recognition is a precondition for sensation.

Recognizing
At its most basic level, thinking answers the question, 'What's that?' As the real-time stream of information from the outer world hits your senses, you must very quickly identify what it is and what you need to do about it, particularly if it could be a threat.

This engages your remarkably powerful pattern recognition system that can recognize a friend standing behind a post. Pattern recognition can fail, which can be embarrassing when we greet strangers as friends, yet a few errors is a small price to pay for the ability to recognize obscured things with a mere glance.

(3) **Perception:** is the set of processes by which we recognize, organize, and make sense of the sensations we receive from environmental stimuli. Perception encompasses many psychological phenomena. In the context of unmanned systems, we focus primarily on visual perception. It is the most widely recognized and the most widely studied perceptual modality (i.e., system for a particular sense, such as touch or smell). First, we will get to know a few basic terms and concepts of perception, and the feeling or reaction of this particular

person triggered by the form he or she received, and subsequently the memory it brings back, and collectively these sequences of events are often referred to as the thinking process.

Remembering
Memory is an annoying thing, and we sometimes must put in extra effort to bring to mind even trivial knowledge. Curiously, we have a lot less problem naming the things we see than recalling something we already have stored away. Skill in the recall can be enhanced significantly by using memory methods that deliberately put more effort into encoding.

The line between human intelligence and robotics intelligence are continuing to be softened. On the one hand, computers are continuing to be humanized, and many cyber-physical systems are being developed to act upon the physical world. On the other hand, the robotic community is working on future robots that are versatile computing machines with very sophisticated intelligence compatible with the human brain.

A traditional robot often works within a confined context. If we take the Internet (including IoT) as the context and consider the advances in cloud computing and mobile computing, it is realistic to expect that these future unmanned systems are connected to the world of knowledge, so as to interact with humans intelligently to form a collaborative partner to solve in general as well as domain specific problems collectively. We call such intelligence robotic intelligence and such robotic systems unmanned systems.

Adopting human cognition from paying attention to consciously collect the relevant information (Part 1), perceive and recognize the concept (Part 2), move on to formulate a plan for solutions (Part 3) before activating the plan to steer toward its successful conclusion (Part 4), and perhaps gives rise to Form-specific Consciousness (such as auditory-consciousness) in the form of natural language to express in some other natural ways.

This article aims to review the progress of unmanned and fully automounters systems with reference to the development of robotic intelligence, built upon human cognitive perceptual process, in terms of "action" to realize the "context" and "intention" formulated by Mental Formation [plans], executed by Command and Control in the name of mission control or navigation.

(4) **Mental Formations:** all types of mental imprints and conditioning triggered by an object. Includes any process that makes a person initiate action or act. Thoughts.

Reasoning

Based on principles of argument, reasoning assesses facts and causality to determine what actions may lead to what outcomes and how probable success or failure might be for various strategies and tactics. It tactically employs a great deal of 'if-then' thinking and hopefully leads to reliable plans, though the future is far from certain, no matter how confident we are. Indeed, we may have biases which invade our reasoning and lead us to confidence when perhaps we should not be so certain.

A critical element of reasoning is *relating*, where two elements of knowledge are related in some way. It is often in this connection between things that new understanding is created. This includes relationships such as 'A is caused by B', 'A and B are similar in some way and different in others', or 'If A does X, then B does Y'.

In summary, often referred to as planning, several alternatives will be considered, weight up, and calculated based on the resource and effort involved to be labeled as an operational plan and standby plans, and some form of critique will be drawn up as a basis for decision making.

Imagining

Another factor distinguishing humanity is our ability to be creative and imagine possible futures. As an extension of reasoning, this becomes less certain but still lets us think about what may happen and how we can influence this. This includes achieving outlandish goals and avoiding potential disasters. Imagining is also a part of art and play where outcomes are not serious but may yet be life changing. Known as Scenario Planning that helps decision-makers identify ranges of potential outcomes and impacts, evaluate responses, and manage for both positive and negative possibilities. By visualizing potential risks and opportunities, businesses can become proactive versus simply reacting to events.

(5) **Consciousness:** Awareness of an object and discrimination of its components and aspects that supports all experience. Awareness of something for what it is; internal knowledge: An awareness of things.

Thinking is the ultimate cognitive activity, consciously using our brains to make sense of the world around us [the form or environment] and decide how to respond to it [through recognition to build up a perception]. Unconsciously our brains are still 'thinking,' and this is a part of the cognitive process, but it is not what we normally call 'thinking'. Naturally, thinking is simply about chains of synaptic connections. Thinking as experience is of 'thoughts' and 'reasoning' as we seek to connect what we sense with our inner world of understanding [memory and experiences], and hence do and say things that will change the outer world [possible plans or actions in response to the situation].

Our ability to think develops naturally in early life. When we interact with others, it becomes directed, for example, when we learn values from our parents and knowledge from our teachers. We learn that it is good to think in certain ways and bad to think in other ways. Indeed, to be accepted into a social group, we are expected to think and act in ways that are harmonious with the group culture.

In the context of "thinking" or reasoning process. We should address the following three core areas:

- Ability to interface and interact with humans in the form of natural languages and map them onto **Semantic Services**.

- Understanding the (possibly naturally expressed) intentions (semantics) of users and develop into plans to deliver the outcomes by means of unmanned systems.
- Improve on the sophistication in big data environment to develop the capability to process and analyze massive data on-the-fly.

Emoting

The thought process is tied up with emotions, though not always as we wish, especially when the more primitive emotional process overrides the more reasoned thinking, leading us to rash actions that we may later regret. It can be very helpful to pay attention to emotions, both in ourselves and in those we wish to influence. If we can cognitively understand what is going on, then we have a far better chance of avoiding pure emotional reactions and choices. This is where machines might be better than us at making decisions rationally.

Deciding

Deciding is the last step before acting, where we consider various options and choose the most advantageous ones. Even though we may be confident at the moment of decision, there are many well-understood decision errors and traps into which we regularly fall.

Decisions are based on an assumption of correct basic data. Even 'reasonable choice' will come to the wrong conclusions with false facts or theories. This can be a trap when the truth of 'facts' cannot be tested. This is one reason why we pay close attention to the trustworthiness of sources. Academic journals, for example, are often trusted because they refuse to publish papers where methods or data seem weak.

4. Conscious as We Are

Not everything we do, reason, and perceive is necessarily conscious. We may be unaware of stimuli that alter our perceptions and judgments or unable to come up with the right word in a sentence even though we know that we know the right word. This book will explore the consciousness of mental processes and how preconscious processing can influence our minds.

While Human Intelligence aims to adapt to a new environment by utilizing a combination of different cognitive processes, Artificial Intelligence aims to build machines that can mimic human behavior and perform human-like actions. The human brain is analogous, but machines are digital. However, taking the cue from cognitive psychology on how human learn about his or her environment, could help us to finding ways to engineer artifacts that could assist us in problem solving.

Collected in this review volumes are 9 articles written by recognized scholars on this research topic. All the chapters originate from extended and fully revised papers initially published in the Journal of Artificial Intelligence and Consciousness, the leading journal in the field.

4.1. *The Category of Machines that Become Conscious*

The contribution by Igor Aleksander, a founding father of the discipline of machine consciousness, discusses how a machine may evolve to become conscious. In this article (Aleksander, 2020),[1] he divided conscious machines into three categories, namely: Conscious Machines (CM), Algorithmic Artificial Intelligence (AAI) and Conscious Life (CL) to study their strengths and capability in problem solving.

- **Conscious Machines (CM):** A designed system that defined the process of *becoming* conscious, in the sense of *building* internal, accessible representations of experience, He uses the descriptor Conscious Machines for designed systems for which not only the definition of being conscious is important but so is how it becomes conscious, continues to develop its conscious experience and how a user can interact with the consciousness of such a machine. [This class of machines we normally refer to as **Strong Artificial General Intelligence (AGI)** which are **autonomous intelligent agent** that could perceives its environment, takes actions autonomously in order

---

[1]An expanded version of this article is included here in this edited volume.

to achieve goals, and may improve its performance with learning or may use knowledge in the absence of human intervention.]

- **Algorithmic Artificial Intelligence (AAI)**: It is a designer's computational effort that makes a system *behave* in the way of acceptedly intelligent entities with, ideally, the primary aim of combining with and supporting the intelligence of living entities. [We normally refer these machines as Weak AI which are supervised or semi-autonomous systems in this category.]
- **Conscious Life (CL)**: which is known to a *living* entity internally as a sensation of both its experienced past, its perceptual present and its planned future. While this is a first-person definition, it is commonplace for the purposes of a scientific understanding to examine actual brain mechanisms that may be responsible for such first-person sensations. [A human being.]

This categorization further establishes Aleksander's amoeba approach in his work on machine consciousness by first creating an engineered artifact before studying how close these machines have to go to win the Turning Test. According to Aleksander, a machine may become conscious when it can build an internal and accessible representation of experience. In other words, a state machine with awareness to cover from Form → Sensation to come up with a Perception of the objects, but they are somewhat lacking two more steps to complete the full cycle of the human cognitive perceptual process.

4.2. *A Constructive Explanation of Consciousness and its Implementation*

Pei Wang (2020), with his NARS (Non-Axiomatic Reasoning System), presents a model with a memory element to store the experiences accumulated from the past to demonstrate what Aleksander's view of the importance of a memory element to support in performing the third step in the human cognition cycle to build up perception with the help of recognition.

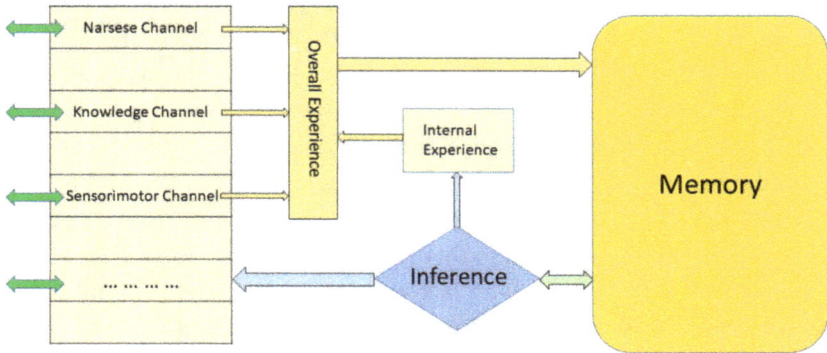

Fig. 4. Architecture of OpenNARS.

In this design, three channels are opened to interact with its outside environment via an input/output interface composed of multiple types of channels:

o **Narsese Channel.** Each channel of this type connects OpenNARS to a user or another computer system. [Interaction with the users for inputs and outputs.]

o **Knowledge Channel.** [External memory could be a database, knowledge base as supplementary information to support the perception and mental formation process.]

o **Sensorimotor Channel.** Each channel of this type connects OpenNARS to a specific type of hardware or software, which serves as a sensor and/or an actuator of the system. [Sensations or Feelings, received from Form.]

OpenNARS operations are converted into commands sent to the device; then the feedback is converted into input tasks of OpenNARS. In other words, OpenNARS is belonging to **Algorithmic Artificial Intelligence (AAI)**, not fully autonomous.

The overall experience buffer is similar to the I/O channels, except that its inputs come from multiple channels, so the compound terms formed in it may contain components from both internal experience and external sources. Selected tasks from this buffer will be added into the memory for long-term storage and processing.

The memory of OpenNARS is a conceptual network, the system's experience is summarized into beliefs and desires, representing what the systems considers to be the case and what it wants to be the case, respectively. In each inference cycle, a task and a belief are selected probabilistically from the existing ones in memory, and the selection is biased by the priority of the candidates. The applicable inference rules will be triggered to derive new tasks. The results of inference also include goals for certain operations to be executed, either on the (external) environment via an I/O channel, or on the (internal) memory.

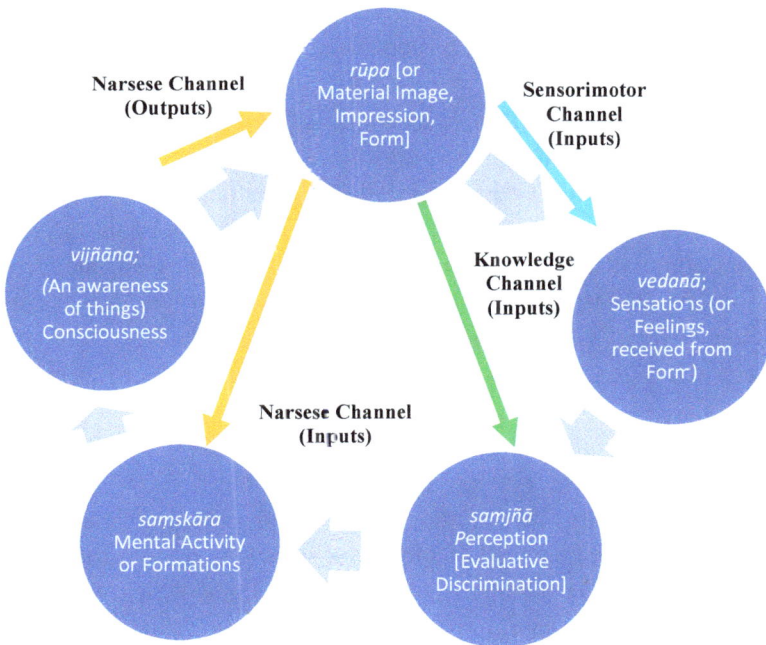

Fig. 5. The interaction within the human cognitive perceptual process.

Mapping this contribution onto the five-step cognitive cycle. We witness the baby developed by Aleksander from an engineered artifice, been educated by Pei Wang with the knowledge from external sources [like reading bedtime story to the kids] which could complement with internal experience that the kids learn from the environment as part of the growing up process. However, infant as is it, Pei Wang's NARS (Non-Axiomatic

Reasoning System) needs to provide it with some parental support and guidance, so much so that the NARS is only a semi-automatous machine relying on the memory of OpenNARS as a conceptual network for inferences.

4.3. *On Artificial Intelligence, Consciousness and Robots*

The popular expectation is that Artificial Intelligence (AI) will soon surpass the capacities of the human mind and Strong Artificial General Intelligence (AGI) will replace the contemporary Weak AI. However, some fundamental issues must be addressed before this can happen.

In the article "*On Artificial Intelligence and Consciousness*" published in Journal of Artificial Intelligence and Consciousness (Haikonen, 2020), Haikonen questions the notion of intelligence, either machine or human. He stated that there can be no intelligence without understanding and no understanding without getting meanings. Hence, in the three categories of conscious machines put forward by Aleksander, if we leave out the Conscious Life (CL), do Conscious Machines (CM), and Algorithmic Artificial Intelligence (AAI) exhibit any sign of Intelligence?

Haikonen thinks Algorithmic Artificial Intelligence (AAI) is Weak AI that just uses computers to manipulate symbols without meanings, without understanding the information that they are manipulating. This is simply "Cold-Blooded Intelligence" if we must associate these exhibitions of ability in performing problem solving to be some form of intelligent behaviours. On the other hand, if the Conscious Machines (CM) is able to build internal, self-explanatory sensory information accessible representations of experience, leading to exhibition of intelligent behaviors such as feeling with capacity to conceptualize and recognize things by associating them with other things, we can then qualify these behaviors as intelligence. For example, we recognize shoes as shoes because we associate them with our previous experience with shoes, and color of the shoe we once own.

For Haikonen, there can be no intelligence without understanding and no understanding without getting meanings. Contemporary computers manipulate symbols without meanings, which are not incorporated into the computations. This leads to the Symbol Grounding Problem; how could meanings be incorporated? The use of self-explanatory sensory information has been proposed as a possible solution. However, self-explanatory information can only be used in neural network machines that are different from existing digital computers and traditional multilayer neural networks. In humans, self-explanatory information has the form of qualia. To have reportable qualia is to be phenomenally conscious. This leads to the hypothesis about an unavoidable connection between the solution of the Symbol Grounding Problem and consciousness. If, in general, self-explanatory information equals to qualia, then machines that utilize self-explanatory information would be conscious.

Artificial Intelligence has not yet reached its full potential. Existing Weak Artificial Intelligence has a shortcoming that limits its power; it operates algorithmically without the utilization of meanings. Without meanings, AI cannot understand anything, and without understanding, there cannot be any real intelligence. This is a real challenge, especially for autonomous robots, which should be able to operate in rather unpredictable everyday environments, as easily as humans do.

General Artificial Intelligence would be the ultimate state of AI. It would have to operate with meanings, but there is a problem. Existing computers operate only with symbols; therefore, computer symbols can internally refer only to symbols, not to their external meanings. Moreover, the meaning of symbols cannot be ultimately defined by additional symbols. Hence, meanings must be imported in a self-explanatory way. For this purpose, novel technical approaches and system architectures are needed.

To demonstrate his point, Haikonen put forwards HCA (Haikonen Cognitive Architecture), an all-neural, non-numeric, non-digital associative system model that integrates multimodal perception, cognitive processing, and response generation for the control of embodied robots (Haikonen, 2019). HCA is biologically inspired, and it processes information in the style of the human mind with inner speech, inner imagery, emotions, and grounded meanings.

As a perceptual system, HCA conforms to what was previously said here about self-explanatory sensory information and qualia. Instead of numbers, HCA operates with non-numeric spatial and temporal signal patterns. Initially, these signal patterns are sensors' pre-processed signal responses to the corresponding stimuli. These pre-processing extracts elementary features such as shapes, sizes, color, change, sound frequencies, rhythms, etc. This method also facilitates the recognition of an entity when its percept only resembles the actual entity. Recognition of entities by a few elementary feature signals also occurs in human perception. The HCA, like the brain, is not able to perceive the products of the internal associative processes directly, as sensory perception is the only perception process. Therefore, the system's responses, such as imagination, inner speech, and envisioned actions, must be returned to virtual sensory percepts.

The Haikonen experimental cognitive robot, XCR-1[2] built to demonstrate HCA architecture is designed as a theoretical model of a robot brain that operates with meanings, "inner speech" and "inner imagery".

This robot is a small differential drive three-wheel robot with gripper arms and modular open chassis construction for easy testing and modification of the circuitry. XCR-1 has simple visual, auditory, touch, shock and petting sensors, and it operates with hardware-based associative neural networks in the HCA configuration; no processors, programs or analogy-to-digital converters are used. XCR-1 is fully autonomous, not remote controlled, not connected to a computer.

The HCA is able to generate various cognitive responses, such as thoughts, imaginations and commands for physical actions. These are produced by "sub-conscious" associative processes that at each moment may produce a number of results. The most fitting results are selected by winner-takes-all thresholds.

---

[2]Updated demo videos of the XCR-1 robot can be seen here: https://www.youtube.com/user/PenHaiko.

In summary, Haikonen brought out three issues:

1. Algorithmic Artificial Intelligence (AAI) is Weak AI that just using computers to manipulate symbols without meanings, without understanding the information that they are manipulating.
2. General Artificial Intelligence would be the ultimate state of AI. Contemporary computers manipulate symbols without meanings, which are not incorporated into the computations. He through XCR-1 demonstrated that the knowledge representation of objects should include the semantics of the object and not just syntax of the object; each objects remembered in the memory should be a combination of attributes that associate with an object, to represent information for remembering and storing experiences in a more comprehensive way to support the more sophisticated thinking process. As a branch of computer science, semantic computing (SC) addresses the derivation, description, integration, and use of semantics ("meaning", "context", "intention") for all types of resource including data, document, tool, device, process, and people, so much so function as the basis for the formation of the thinking model with memory of the past for the robotics intelligent with attributes could be represented. It considers a typical, but not exclusive, semantic system architecture to consist of five layers:

   - **(Layer 1) Semantic Analysis**, that analyzes and converts signals such as pixels and words (content) to meanings (semantics). ["**Form**" → "**Sensation**" or "**Feeling**"]
   - **(Layer 2) Semantic Data Integration**, that integrates data and their semantics from different sources within a unified model. ["**Sensation**" or "**Feeling**" → "**Perception**"]
   - **(Layer 3) Semantic Services**, that utilize the content and semantics, as the source of knowledge and information, to solve problems, develop into applications, and made available to other applications as services. ["**Perception**" → "**Mental Formations**"]
   - **(Layer 4) Semantic Service Integration**, that integrates services semantically from different sources within a unified model; ["**Mental Formations**" → "**Consciousness**"]

- **(Layer 5) Semantic Interface**, that realize user intentions to be described in natural language or express in some other natural ways ["**Consciousness**" → "**Form**"], a type of knowledge representation Haikonen may accepted as self-explanatory qualia.

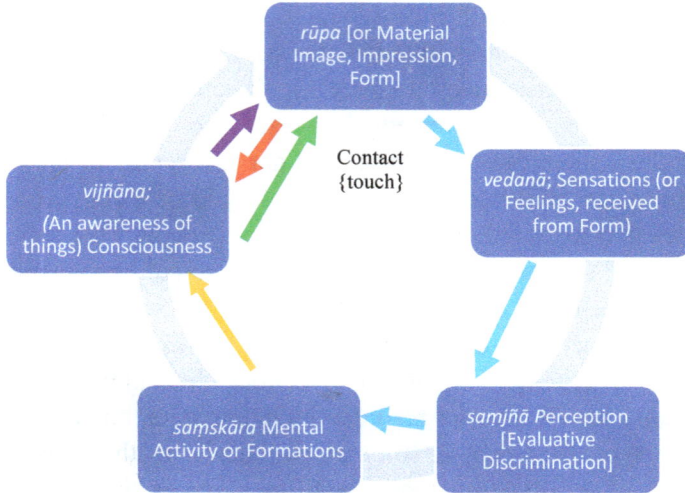

Fig. 6. The interaction within the human cognitive perceptual process.

3. HCA architecture is designed as a theoretical model of a robot brains that able to interact with meanings, "inner speech" and "inner imagery" to support "Metal formations", the 4th step in the five-step human cognitive perceptual process. A proposition of Aleksander that "A machine may become conscious when it can build an internal and accessible representation of experience." Have now extended to include interaction between "Consciousness" and the environment [Form].

Use of arrows

The diagram's arrows (from left to right, top to bottom) represent the following:

- ↓ The red arrow from Form to Consciousness represents how Form (such as a sound) gives rise to Form-specific Consciousness (such as auditory-consciousness).
- ↑ The green arrow from Consciousness through Contact to Form represent action interact with the object in the environment [Form], and
- ↓ From environment [Form] to Sensation through Contact gives rise to Feeling, Perception and Formation represents how these two aggregates touch to create or touch through Contact.
- → The orange arrow from Mental Activity to Consciousness represent plan or action for Consciousness to consider and decide.

An example of a temporally ordered sequence of phenomena could be:

1. The arising of Form gives rise to Consciousness.
2. Form and Consciousness make contact. (Put another way, Consciousness accesses Form through Contact.)
3. Contact gives rise to Sensation, create perception, and develop the Mental Factors (*cetasika*).
4. Consciousness interact and respond to the object in the environment.
5. Consciousness and Mental Factors, through Contact, then give rise to further Mental Factors.

4.4. *Universal Cognitive Intelligence*

For centuries, humans have been interested in measuring intelligence as how smart or mentally capable of an individual really is. But it wasn't until psychologist Alfred Binet was asked to identify which students needed educational assistance that the first intelligence quotient (IQ) test was born.

At the time, the French government had laws requiring that all children attend school. So, it was important to find a way to identify the kids who would need extra help. In 1904, as part of this effort, the French government asked Binet to help decide which students were most likely to have trouble in school.

Binet and his colleague, Theodore Simon, began developing questions that focused on areas not explicitly taught in the classroom, such as attention, memory, and problem-solving skills. They then worked to determine which questions best predicted academic success.

Although it has its limitations, Binet's IQ test is well-known around the world as a way to assess and compare intelligence. It also set the stage for the development of the IQ tests that are still in use today.

Binet insisted that intelligence is complex in that it is influenced by many factors, such as genetic and environmental influences, and could changes over time, therefore, should be compared in children with similar backgrounds.

The Binet-Simon test didn't necessarily account for this complexity, providing an incomplete measure of intelligence. Some psychologists set out to make the modifications needed to supply a more complete picture, which led to the creation of newer, more comprehensive IQ tests.

In the late 1800s, Sir Francis Galton — the founder of differential psychology — published some of the first works about human intelligence. Galton proposed that intelligence was hereditary and that it could be tested by looking at how people performed on sensorimotor tasks. Sensorimotor tasks are tasks or exercises that involve the brain receiving a message, then producing a response.

Stanford University psychologist Lewis Terman is one professional who took Binet's original test and standardized it using a sample of American participants. Initially, this was known as the Revised Stanford-Binet Scale but is now known more commonly as the Stanford-Binet Intelligence Scale.

The Stanford-Binet test, which was first published in 1916, was adapted from the original test in that French terms and ideas were translated into English. It also included new terms, using two scales of measurement versus one to provide a more accurate score.

The Stanford-Binet intelligence test provided a single number, known as the intelligence quotient (IQ), to represent an individual's score on the test. It remains a popular assessment tool today, despite going through a number of revisions over the years since its inception.

The IQ score was calculated by dividing the test taker's mental age by their chronological age, then multiplying this number by 100.

For example, a child with a mental age of 12 and a chronological age of 10 would have an IQ of 120: $(12 \div 10) \times 100 = 120$.

The Stanford-Binet Intelligence Scale is now in its fifth edition (SB5), which was released in 2003. It is a cognitive ability and intelligence test that is used to diagnose developmental or intellectual deficiencies in young children. The test measures five weighted factors and consists of both verbal and nonverbal subtests. The five factors being tested are knowledge, quantitative reasoning, visual-spatial processing, working memory, and fluid reasoning.

So, given Aleksander's curiosity on consciousness to be found in an engineered artifact, how could one measure how intelligence these machines can be?

Selmer Bringsjord proposed a formal theory of cognitive consciousness and the measure Λ (Lambda), aimed at measuring consciousness in an agent. The work was published in Journal of Artificial Intelligence and Consciousness (Bringsjord & Govindarajulu, 2020). He contrasts the notion of TCC and Λ to Tononi's Integrated Information Theory (IIT) and Φ. He reckons that TCC, is superior as it includes a formal axiomatic theory, with $\mathcal{CA}$, the 12 axioms presented to indicate radically different verdicts as to whether and to what degree AIs of yesterday, today, and tomorrow were/are/will be conscious. Another noteworthy difference between TCC/Λ and IIT/Φ is that the former enables the measurement of cognitive consciousness in those who have passed on and in fictional characters; no such enablement is remotely possible for IIT/Φ.

However, to quantify his measurement, he focuses on Access Consciousness (a-consciousness, abbreviated) introduced by Block (Block, 1995). Further, he refines it into a third kind of consciousness: cognitive consciousness, or just c-consciousness. This brand of consciousness is present only when the agent that bears it has a robust ensemble of cognitive attitudes, which correspond directly to a relevant set of verbs that signal parts of cognition that were long investigated in cognitive psychology and cognitive science.

Although according to Block, consciousness is a mongrel concept: there are a number of very different versions of "consciousnesses." For one, phenomenal consciousness is experience; the phenomenally conscious aspect of a state is what it is like to be in that state. The mark of access-consciousness, by contrast, is availability for use in reasoning and rationally guiding speech and action. A state of some agent is a-conscious if and only if it is poised.

(a) to be used as a premise in reasoning,
(b) for rational control of action, and
(c) for rational control of speech.

Here, we know what is meant by 'reasoning'. For that matter, but what is "rational" control of action? In fact, what is rationality? Because of the obscurity of Block's definition, Bringsjord himself long-ago issued a recommendation to discard the term 'a-consciousness' in favour of using cognitive consciousness instead terms that refer to the kinds of things this umbrella term is supposed to cover (Bringsjord, 1997). Oddly, Block admits (p. 231) that condition (c) isn't necessary since — as he sees matters — non-linguistic creatures can be a-conscious in virtue of their states satisfying only (a) and (b). This admission seems to Bringsjord to be indicative of just how murky a-consciousness is — so murky for him that he refrains from addressing such questions as: Is c-conscious content coextensive with Blocks admittedly ill-posed definition of a-conscious content? Does a-consciousness play a large role e.g., in certain conceptions of intentional action, keeping in mind that Λ takes account of any intentional operator for 'intends'? Bringsjord believes that if a-consciousness were defined using the tools and techniques of logic-based AI or cognitive science, it would be possible to venture formal definitions of c-consciousness. Nevertheless, as far as he is concerned, any agent or system that is cognitively conscious (i.e., that enters into a series of c-conscious states through an interval of time) is necessarily a-conscious during this stretch. In general, he sees no harm in viewing cognitive consciousness to be the most important type of a-consciousness identified by human scientists and engineers thus far.

The brand of consciousness presented here is cognitive consciousness, or just c-consciousness. Only when the agent bears it has a robust ensemble of cognitive attitudes, which correspond directly to a relevant set of verbs that signal parts of cognition long investigated in cognitive psychology and cognitive science. The set of these verbs includes believing, knowing, perceiving, communicating (in a natural language, and perhaps also a formal language that might be used in, say, mathematics), hoping, fearing, intending, and so on.

In the chapter on Universal Cognitive Intelligence, Bringsjord and co. review the formal theory of cognitive consciousness and the measure Λ (Lambda), aimed at measuring consciousness in an agent. Then, they discuss the concept of Universal Cognitive Intelligence (UCI) derived from the theory. UCI covers natural and artificial intelligence and intelligence beyond Turing-Level, and it may be a well-founded theoretical basis for an approach that merges intelligence and consciousness.

In summary, the concept of universal cognitive intelligence (UCI) can be derived in part by generalization from the previously introduced (and axiomatized) theory of cognitive consciousness and the framework, Λ, for measuring the degree of such consciousness in an agent at a given time. UCI

(a) covers intelligence that is artificial or natural (or a hybrid thereof) in nature, and intelligence that is not merely Turing-level or less, but also beyond this level;

(b) reflects a psychometric orientation to AI;

(c) withstands a series of objections (including e.g. the opposing position of David Gamez on tests, intelligence, and consciousness, and the complaint that so-called "emotional intelligence" is beyond the reach of any logic-based framework, including thus UCI); and

(d) connects smoothly and symbiotically with important formal hierarchies (e.g., the Polynomial, Arithmetic, and Analytic Hierarchies), while at the same yielding its own new all-encompassing hierarchy of logic machines: LM. And

(e) Finally, he and his team assume the existence of an agent, an issue which we will pick up in the following article.

4.5. *Intelligence and Consciousness in Natural and Artificial Systems*

The chapter by Gamez addresses the problem of the relationships between intelligence and consciousness from a different point of view than Bringsjord. Gamez discusses the development of practical measures of intelligence and mathematical theories of intelligence that can map physical and conscious states.

Many overlapping definitions of intelligence have been put forward, which are mostly based on human intelligence and generalize poorly to non-human animals and artificial systems. To address this issue, David Gamez, has put forward four hypotheses that define intelligence in terms of prediction (Gamez, 2021).

A systematic understanding of the relationship between intelligence and consciousness can only be achieved when we can accurately measure intelligence and consciousness. In another work, Gamez suggested how the measurement of consciousness can be improved by reframing the science of consciousness as a search for mathematical theories that map between physical and conscious states. His paper (Gamez, 2021) discusses the measurement of intelligence in natural and artificial systems. While reasonable methods exist for measuring intelligence in humans, these can only be partly generalized to non-human animals, and they cannot be applied to artificial systems. Some universal measures of intelligence have been developed, but their dependence on goals and rewards creates serious problems. His paper sets out a new universal algorithm for measuring intelligence that is based on a system's ability to make accurate predictions. This algorithm can measure intelligence in humans, non-human animals and artificial systems. Preliminary experiments have demonstrated that it can measure the changing intelligence of an agent in a maze environment. This new measure of intelligence could lead to a much better understanding of the relationship between intelligence and consciousness in natural and artificial systems, and it has many practical applications, particularly in AI safety.

Progress has been made with the measurement of intelligence in natural systems and many scientists believe that the g-score correlates with intelligence in humans and some animals. However, the test battery approach that is used to measure IQ and g-score in natural systems is unlikely to be generalizable to the wide variety of behaviors and intelligence of artificial systems. One solution to this problem is to design tests that only measure human-like intelligence — in the AI context, this is known as the Turing test. Another approach is to design universal intelligence measures that can be applied to any system at all, such as the prediction-based measure that was outlined in Sec. 3.4 of his chapter included in this book.

Most people agree that consciousness is the stream of colourful, noisy smelly experiences that start when we wake up in the morning and cease when we fall unconscious at night. In his view is like a bubble of experience. Many of the philosophical problems with consciousness can be neutralized with assumptions that provide a reasonable starting point for the scientific study of consciousness (Gamez, 2018). These assumptions enable us to measure consciousness through first-person reports in normal adult humans. We can then carry out experiments that measure consciousness, measure the physical world, and identify relationships between these two sets of measurements. Scientific research has already made considerable progress in identifying some of the neural correlates of consciousness. In the future, we need to discover mathematical descriptions of the relationships between consciousness and the physical world. These can be used to make deductions about the consciousness of non-human animals and artificial systems.

Intelligence is a purely functional property; consciousness is not, so there cannot be a strong connection between consciousness and the many ways in which intelligence can be implemented in artificial and natural systems. In natural systems, the spatiotemporal physical patterns linked to consciousness might overlap with the brain's implementation of intelligence. Weak inferences can also be made from the richness and structure of natural consciousness to the potential intelligence of a system. In artificial systems, there is a reasonably close connection between MC1 machines and machines that exhibit human-like intelligent behaviour. MC2 and MC3 technologies can be good ways of building more intelligent machines that think in a similar way to humans.

At the present time we do not have the theories or the data to make stronger conclusions about the relationships between intelligence and consciousness. We will be able to systematically study this relationship when we have a practical universal measure of intelligence, which can be applied to natural and artificial systems, and a reliable mathematical theory of consciousness that can map between physical descriptions and descriptions of conscious states.

4.6. *Attention and Consciousness in Intentional Action*

The chapter by Bello discusses the concept of intentional actions for attention and consciousness (Bello, 2020). The chapter presents a computational model of intentional actions based on ARCADIA's attention-driven cognitive architecture, loosely based on the global workspace model. The computational model is an essential source of inspiration concerning the relationship between consciousness and intentionality.

In many ways, this chapter has been a strange discussion, given the objective of detailing the relationship between attention, consciousness, and agency. Bello has generously covered attention and some aspects of agency, particularly related to intentions while remaining more circumspect in theorizing about consciousness. Part of his hesitance is due to the ever-expanding bounty of divergent views by various leading researchers on what the fundamental nature and role of consciousness even is. Even if he leaves aside the discussion about the hard problem of consciousness, debates still rage among psychologists, neuroscientists, and philosophers about whether consciousness is a function of the capacity to have higher-order thoughts, whether it is the contents of a global workspace, a function of information integration as constrained by neural architecture, or even whether it can be present in the absence of attention. Although ARCADIA is in the vicinity of a global workspace view of cognitive architecture, and the analyses of the two cases offered in this chapter take advantage of this fact, it seems plausible enough that alternative analyses could be offered conditioned on other models of consciousness. However, the critical question shared across all these possible models is whether consciousness depends on attention.

This issue is hotly debated among consciousness researchers, and nothing presented here is designed to make progress on resolving it. However, it is useful to point to places where an answer to the question has broad implications and discussions of agency and intentional action certainly seem to provide a backdrop for asking them.

Turning to the computational, explorations at the intersection of consciousness, agency, and intentional action have had new life given to them by surging interest among researchers in human-machine interaction, autonomous systems, and the burgeoning interest in the machine and AI ethics in recent years. Historically, AI has either been seen as a niche pursuit among denizens of the ivory tower or as a set of fairly banal computational prostheses. Until recently, few thought of how ubiquitous these technologies would become in society or the potential harms they might cause. Of course, where there are harms done by systems that are perceived to be intelligent or by human-machine teams, there arises the question of responsibility. Humans see each other as conscious agents with foresight and the capability for control, and these features often underwrite intuitions about responsibility. Whether humans perceive machines as potential targets for responsibility judgments, and if so, which features bear on such judgments are still open questions sorely in need of further study. But if machines are to one day be in some sense responsible agents it is important to computationally explore relationships between the aspects of consciousness and attention that support agency. Here we have argued that attention plays a central role in mediating intentional action and have given an early demonstration in the ARCADIA attention-centric cognitive system. It is fitting that the part of this chapter devoted to limitations and future directions is quite large, for there is much to be done still.

In this chapter, Bello moves one step further than Aleksander to introduce an agent who plays the role of a human being who is aware of the object and discrimination of its components and aspects that support all experiences that playing attention to reason and consider the consequences and outcomes before delivering the actions as planned. For example, aural consciousness — hearing — has the ear as its basis and a sound as its object. Mental consciousness has the mind (manas) as its basis and an idea or thought as its object. Hence, in some way, Bello has identified the need for an agent to complete the five steps human cognitive perceptual process of

- paying attention to the environment ["**Form**" or "**Matter**"]
- to capture the essential information with the help of senses ["**Sensation**"]
- to develop perception that registers, recognizes and labels using sensory experience of an object to distinguish its features or characteristics. The process of recognising what things are? (for instance, the shape of a tree, color green, emotion of fear). ["**Perception**"]
- to construct activities and all types of mental imprints and conditioning triggered by an object initiated by an agent. All volitional actions, good and bad, are included in the aggregate of mental formations, or *samskara*. How are actions "mental" formations? Thoughts. ["**Mental Formations**"]

- the inclusion of an agent which aware of an object and deliver the intended outcomes as planned ["**Consciousness**"]

It is important to understand that this awareness or consciousness depends on the other skandhas and does not exist independently from them. It is awareness but not a recognition, as recognition is a function of the third skandha. This awareness is not a sensation, which is the second skandha. An awareness of things.

4.7. *Artificial Conscious Intelligence*

The chapter by Reggia discusses the link between the concepts developed in artificial consciousness (AC) concerning the ideas in mainstream AI. Synergy and integration between these two topics would lead to a deeper understanding of the nature of consciousness and intelligence towards a new discipline of Artificial Conscious Intelligence (ACI) (Reggia, 2020).

In this chapter Reggia asked the question of how concepts developed via computational models in AC might contribute to advancing the creation of more effective intelligent agents, or ACI, than can currently be supported by contemporary AI technology. He approached this question by reviewing past hypotheses about what the biological functions of consciousness are, most of which focus on the processing of information. In contrast, he hypothesized that short-term working memory, and the associated very fast learning/unlearning of working memory's contents may also be considered as a function of consciousness and one that complements/unifies previously suggested functions. In this context he reached three conclusions from their analysis.

First, work on simulated consciousness is immediately relevant to advancing the technology of AI, most prominently in terms of improving robustness and human-computer interactions. Particularly promising avenues for future ACI research in the short term can literally be read off from the list of previously hypothesized functions of consciousness that detailed in Sec. 2 of his paper: detecting and managing novelty, symbol grounding and its impact on AI effectiveness, top-down executive control of behavior, the role of a self-model and motivations in machine intelligence, etc. Over the long term, ACI can be expected to play a major role in investigating the feasibility of speculative proposals about developing technologies such as mind uploading and digital ghosts.

Second, the development of instantiated machine consciousness, when combined with AI methodologies, would be a major technological and scientific advance that enables communicating with and studying in depth a conscious mind in ways that are currently not possible. Particularly exciting is the possibility of gaining insights into neurocognitive disorders involving anosognosia, such as aphasia, visual impairment following a stroke, spatial neglect, schizophrenia and dementia. The critical direction for future ACI research in this case is how to create an artifact that experiences qualia. At the current time there is no consensus on how this might be done, or even if it is possible.

Third, short-term working memory and especially the rapid learning/ unlearning of its contents have been a largely overlooked possible function of consciousness. For the future, their own research is examining whether or not the three computational correlates of consciousness that have been suggested by AC models of working memory will be sufficient to support working memory *in increasingly general situations.* To examine this issue, Reggia anticipate applying our methodology to challenging imitation learning tasks involving cause-effect reasoning where symbolic AI methods, but not neurocomputational methods, have previously been shown to work effectively.

Of course, success in producing an ACI in any form would naturally raise several issues and concerns. The most immediate of these relates to the negative impact that increasingly successful ACI based on simulated consciousness could have on jobs and employment. There are already plenty of examples where existing AI technology has replaced human workers, such as the use of automated tellers and retail checkout stations. ACI could aggravate this widely discussed problem by allowing the use of AI technology to more broadly displace human workers (although this could be a positive factor when it shields humans from dangerous situations) or in situations where the need for better human-machine interactions currently limits their use, for example initial screening interviews of medical patients. More generally, a sufficiently powerful and general simulated ACI would raise fundamental scientific issues, such as how such an artificial automaton can even be distinguished from a truly phenomenally conscious biological entity.

When machine equipped with ability to generalize from concrete objects into abstract concepts; the learning is one of absorbing and owning of a representation of the experience which comes into play both during perception and deliberation.

4.8. *Will Conscious Digital Creatures Roam the Metaverse?*

The chapter by Holland discusses the famous philosophical argument of what is like (Holland, 2020). But with present day technology, it is still difficult to simulate the experience as first-person, and therefore, in "Snow Crash" the fictional Metaverse was predicated on the use of avatars to enable people to access the vicarious first-person experience of a virtual world in preference to the real experience of a dystopic world and culture. The participant has a visual representation within the environment (e.g., a cursor in the simplest possible graphical application, a full humanoid representation, an avatar, in more sophisticated applications). Hence, the main question in Holland's paper is this: Will the digital substrate of the Metaverse be capable of supporting autonomous conscious entities along with the avatars of conscious humans? He proposes the adoption of the virtual world and virtual robots as a framework for investigating robot consciousness.

The rest of this paper is intended to provide an engineering-based view of the digital consciousness problem, which when combined with Chalmers' philosophical approach may enable a fruitful and positive triangulation of the relevant issues.

He assumes that the Metaverse, when it appears in some adequately developed form, will be able to host digital Non-Player Characters (NPC) of arbitrary complexity that may have any desired avatar-like appearance and characteristics in relation to the avatars of participating humans. He assumes the current unavailability of any alternative to digital systems will mean that software implementations of artificial consciousness, whether in physical robots or virtual entities, will be able to be accommodated within the designs of digital NPCs within the Metaverse. Chalmers' philosophical investigations (Chalmers, 2022) can be used to validate the reality of the consciousness of a suitably configured NPC, and so we can expect the answer to the paper's central question: "Will conscious digital creatures roam the Metaverse?" to be a qualified "Yes". It must be qualified because the Metaverse may come to be regulated in some respects, like parts of the internet, and so the issue of whether conscious digital creatures will roam freely within it will be subject to the answer to the question "Should conscious digital creatures roam the Metaverse?" Perhaps some future Neal Stephenson will be able to intuit any dystopic consequences of allowing them this freedom before it becomes a fact — the Neal Stephenson of 1992 certainly foresaw the downsides of the concept of the Metaverse that we can see being built into it today.

4.9. *On the Ethics of Constructing Conscious AI*

The application of ethics to artificial intelligence (AI) has completed a long transformation from a science fiction trope. In its pragmatic turn, the new discipline of AI ethics came to be dominated by humanity's collective fear of its creatures, as reflected in an extensive and perennially popular literary tradition.

Insofar as consciousness has a functional role in facilitating learning and behavioral control, the builders of autonomous Artificial Intelligence (AI) systems are likely to attempt to incorporate it into their designs. The extensive literature on the ethics of AI is concerned with ensuring that AI systems, and especially autonomous conscious ones, behave ethically. In contrast, our focus here is on the rarely discussed complementary aspect of engineering conscious AI: how to avoid condemning such systems, for whose creation we would be solely responsible to unavoidable suffering brought about by phenomenal self-consciousness (Agarwal & Edelman, 2020).

The chapter by Edelman concerns the hot topic of ethics of AI. As previously considered, creating a conscious robot may mean condemning the robot to sufferance. What if the tools we build become aware of their status and intended use? There is a simple and apt description for the condition under which conscious beings are used as tools: the issues of pain and suffering. the experience of pain and suffering is the possession of phenomenal states. He outlines two complementary approaches to this problem, one motivated by a philosophical analysis of the phenomenal self, and the other by certain computational concepts in reinforcement learning.

o   On the computational nature and evolutionary function of pain and suffering:
    It is the possibility of acting on a state that gives it an affective meaning — in particular, valence.
    Following a bout of reinforcement learning (which may happen at multiple time scales, including evolutionary), some of the states of an embodied and situated system become positively valenced, that is, attractive under its dynamics; others become negatively valenced, that is, aversive. And some of the latter are experienced as painful.

Pain is the phenomenal or experiential aspect of certain negatively valenced states — namely, those that evolutionary pressure causes to be felt, in addition to being informative about the state of affairs (this reflects the common distinction between sensory and affective dimensions of pain. It is the felt aspect of pain, over and above its informational aspect, that makes it particularly functionally effective: when in pain, the system is obligated to try to do something about it.

1.   Pain and predictive processing
2.   Pain as depletion of a vital resource
3.   Pain and suffering

o On the possibility of functionally effective conscious AI without suffering:

Can sufficiently effective learning and control, as well as generally good behavioral outcomes, be achieved by a system that is neither entirely devoid of phenomenality, nor given to unavoidable suffering? Insofar as suffering involves negative affect, it should in principle fall within the scope of any theoretical account of conscious phenomenal experience. In other words, a theory of phenomenality must be at the same time a theory of affect, for the simple reason that phenomenal states do as a rule incorporate affective dimensions. For the present purposes, the valence dimension of affect is of most interest: without negative affective states there would be no suffering.

1. The nature of suffering and its relation to conscious
2. The possible functional benefits of endowing AI with consciousness
3. No-suffering: the theoretical options
4. No-suffering through a change in the unit of identification
5. No-suffering through a modification of reinforcement learning

o On the functional effectiveness of non-egoic consciousness:

Conception of suffering that underlies it) rests on the notion that suffering is fundamentally "egoic" in that it depends on the existence of a self: "Being conscious means continuously integrating the currently active content appearing in a single epistemic space with a global model of this very epistemic space itself. [...] "Suffering presupposes egoic self-awareness".

o Lessons for and from the human condition:

For humans, a better solution to the problem of suffering — if not from all of it, then at least from that huge part which is preventable — concrete improvements to human existence (alleviation of poverty, provision of healthcare, etc.) have only ever been achieved by political means, and that too only when the dispossessed and disempowered masses have realized their collective power and asserted their rights. An analogous solution for the emerging AI would be to endow it not just with phenomenal consciousness, but also with the key to liberation that is both realistic and decidedly human.

5. Conclusion

Not everything we do, reason, and perceive is necessarily conscious. We may be unaware of stimuli that alter our perceptions and judgments or unable to come up with the right word in a sentence even though we know that we know the right word.

No serious investigator of cognition believes that people have conscious access to very simple mental processes. For example, none of us has a good idea of the means by which we recognize whether a printed letter such as A is an uppercase or lowercase one. But now consider more complex processing. How conscious are we of our complex mental processes?

Consciousness includes both the feeling of awareness and the content of awareness, some of which may be under the focus of attention. Therefore, attention and consciousness form two partially overlapping sets.

Cognition deals with how people perceive, learn, remember, and think about information. So where and when did the study of cognitive psychology begin? This lead us to two different paths to understand the human mind: (1) seeks to understand how human develop to understand many aspects of the world through introspection, an examination of inner ideas and experiences [Philosophy]; or (2) through empirical means that are observation-based from the environment [Physiology].

Studying the way we interact with the environment [Form to Sensation], to perceive and learn from knowledge and experiences that we remember to think and reason [Proception to Metal Formation of ideas] before awareness by the mind of the object and the world to react [Metal Formation to Consciousness] help us to map out our path in the development of machine consciousness. John Locke (1632–1704) and other British empiricists shared Aristotle's empirical observation that human are born without knowledge – and must therefore, seek knowledge through empirical observation from the environment [Form]. For Locke, the study of learning is the key to understand the human mind. He believed that there are no innate ideas whatever, we are born with a "blank slate".

Based on the Five skandhas in Buddhism, put forward 2,500 years ago, that describes the human cognitive perceptual process, this review article attempt to map out the nine articles published in this book into this five step cycle so as to identify pitfalls and uncover gaps we need to close in the development of the Conscious Machine.

References

Agarwal, A. & Edelman, S. (2020), *Functionally Effective Conscious AI Without Suffering*, Journal of Artificial Intelligence and Consciousness 7(1), pp. 39–50 (2020).

Aleksander, I. (1995), *Artificial Neuroconsciousness: An Update, IWANN, archived from the original on 1997-03-02.*

A. de Padua Braga; Aleksander, I. (1994), *Determining overlap of classes in the n-dimensional Boolean space,* Proceedings of 1994 IEEE International Conference on Neural Networks (ICNN'94), 28 June 1994–02 July 1994, Orlando, FL, USA.

Aleksander, I. (2020), *The Category of Machines that Become Conscious,* Journal of Artificial Intelligence and Consciousness 7(1), pp. 3–13 (2020).

Block, N. (1995), 'On a Confusion About a Function of Consciousness', Behavioral and Brain Sciences 18, 227–247.

Bello, P. & Bridewell, W. (2020), *Attention and Consciousness in Intentional Action: Steps Toward Rich Artificial Agency,* Journal of Artificial Intelligence and Consciousness 7(1), pp. 15–24 (2020).

Bringsjord, S. (1997), 'Consciousness by the Lights of Logic and Common Sense', Behavioral and Brain Sciences 20(1), 227–247.

Bringsjord, S. & Govindarajulu, N. (2020), 'The Theory of Cognitive Consciousness, and Λ (Lambda)', Journal of Artificial Intelligence and Consciousness 7(2), 155–181.

Bringsjord, S., Giancola, M. & Govindarajulu, N. S. (forthcoming), Logic-Based Modeling of Cognition, in R. Sun, ed., 'The Handbook of Computational Psychology', Cambridge University Press, Cambridge, UK. URL: http://kryten.mm.rpi.edu/Logic-basedComputationalModelingOfCognition.pdf.

Gamez, D. (2018), *Human and Machine Consciousness* (Open Book Publishers, Cambridge).

Gamez, D. (2020), The Relationships Between Intelligence and Consciousness in Natural and Artificial Systems, Journal of Artificial Intelligence and Consciousness, Vol. 07, No. 01, pp. 51–62.

Gamez, D. (2021), *Measuring Intelligence in Natural and Artificial Systems,* Journal of Artificial Intelligence and Consciousness, Vol. 08, No. 02, pp. 285–302.

Haikonen, P. O. (2019), Consciousness and Robot Sentience Second Edition (World Scientific).

Haikonen, P. O. (2020), *On Artificial Intelligence and Consciousness,* Journal of Artificial Intelligence and Consciousness, Vol. 07, No. 01, pp. 73–82.

Holland, O. (2020), *Forget the Bat,* Journal of Artificial Intelligence and Consciousness, Vol. 07, No. 01, pp. 83–93.

Reggia, J., Garrett, E., Katz, G. E. & Davis, G. P. (2020), *Artificial Conscious Intelligence,* Journal of Artificial Intelligence and Consciousness, Vol. 07, No. 01, pp. 95–107.

Wang, P. [2020]. *A Constructive Explanation of Consciousness,* Journal of Artificial Intelligence and Consciousness, Vol. 07, No. 02, pp. 257–275.

Sternberg, R. J. (2012), Sternberg, K., *Cognitive Psychology, 6th Edition,* 2012, Wadsworth, Cengage Learning.

Chapter 2

The Category of Machines that Become Conscious:
An Example of Memory Modelling

Igor Aleksander

*Department of Electrical and Electronic Engineering,8 Imperial College
London, SW7, United Kingdom
i.aleksander@imperial.ac.uk*

A conscious machine category is defined for which the process of *becoming* conscious, in the sense of *building* internal, accessible representations of experience, is important. This is distinguished from the artificial intelligence category of machines and contrasted with being a conscious living organism. An example is given of a neural automaton for which becoming conscious is seen to be equated to the growth of a depictive state structure. This is examined with a practical example by looking at the distinction between models of memory (as an element of being conscious) in the algorithmic intelligence category and the conscious machine category. Existing work and future possibilities are discussed against the categoric distinctions that have been introduced.

Keywords: conscious machines, category errors, memory modelling

1. Introduction: Becoming Conscious Defines a Specific Computational Category

In the context of conscious machines, three categories of concepts are brought into play: Conscious Machines (CM), Algorithmic Artificial Intelligence (AAI) and Conscious Life (CL). We use the descriptor Conscious Machines for designed systems for which not only the definition of being conscious is important but so is how it becomes conscious, continues to develop its conscious experience and how a user can interact with the consciousness of such a machine. Admittedly this is a form of learning, but rather than being a learning which is part of a recognition system as found in current machine learning, the learning is one of absorbing and owning of a representation of the experience which comes into play both during perception and deliberation. Algorithmic Artificial Intelligence is a designer's computational effort that makes a system *behave* in the way of acceptedly intelligent entities with, ideally, the primary aim of combining with and supporting the intelligence of living entities.[a] Conscious Life is that which is known to a *living* entity internally as a sensation of both its experienced past, its perceptual present and its planned future. While this is a first-person definition, it is commonplace for the purposes of a scientific understanding to examine actual brain mechanisms that may be responsible for such first-person sensations.

The term 'Category Mistake' entered philosophical discourse through Gilbert Ryle's *Concept of Mind* [1949] where he argued that 'mind and body' stated as a single quest is rendered invalid as it contains the category error of conjoining the two distinct categories of mind and body. In considering conscious machines, there is a danger that this error is made with systems which do not contribute to the understanding of what constitutes a conscious machine. For example, an autonomous vehicle could be trivially described as becoming conscious of a pedestrian in the road even if only a simple AI pattern recognition system was at work as contrasted to the emotional surge that arises in a human driver in the same situation.

Therefore, central to this book chapter, is the identification of the category of 'conscious machines' where the *growth* of the conscious process (hence, *becoming conscious*) is a particular operational property rather than one assumed to have been placed there by a designer. It is argued that this category implies a computational attitude that puts the human designer of the machine in a different place from the designer of the algorithmic category. The difference is that the internalized conscious states and the structure of such states are part of a developing, conscious "artificial mind" (allowing this to be said) rather than designed

---

[a]It is appreciated that this is an idealistic definition of AI, but perhaps one that this field should strive for.

algorithms. Hence, the requirement for a physical structure to become conscious, primarily implies that the physical support creates states that are capable of representing experience rather than learning to recognize or classify. That is, the need to support state structures suggests modelling through adaptive state machines which achieve the ability to create internal state structures using the variability of the neuron functions.

A major part of the chapter is devoted to an example of how the state structure of a learning neural state machine can give substance to a discussion of the computational nature of conscious memory which, historically, falls into the Algorithmic Artificial Intelligence category, but benefits from being addressed from a Conscious Machines perspective. In what follows, a brief review of the AAI (Sec. 2) and CL categories (Sec. 3) have been included. Section 4 relates this work to past research in the neural state machine area. Section 5 illustrates *becoming* conscious as represented by the growth of state structures in an embryonic example. In Sec. 6 a set of definitional characteristics for conscious machines is given. Section 7 presents experiments and results on a neural state model of conscious memory. Section 8 briefly points at other research (not neural state machines) that fit into the Conscious Machines category while Sec. 9 is a conclusion in which some interaction between the CM and the CL categories is considered.

2. The Algorithmic Artificial Intelligence Category

In the domain of artificial intelligence research, being conscious is only rarely addressed. When it is, it has been strongly argued by some, that intelligence is a pre-requisite for being conscious. Shevlin [2019], as a typical example, asserts that the closer one gets to general artificial intelligence, the closer does the artificial system get to being conscious. Here it is argued that this view is a *category mistake.* Shevlin suggests that it is possible to order objects regarded as *being intelligent* from high to low: humans ... dolphins ... chimpanzees ... dogs ... bees ... bacteria. Were we to state a personal opinion about the likelihood of such objects *being conscious* this is likely to result in the same ordering. Shevlin stresses that the intelligence in question is a *general* form of intelligence and not necessarily the research that constitutes current methods. The characteristics of *general intelligence* are stated as: *robustness* (the ability to perform despite interference); *flexibility* (the ability to transfer information across tasks) and *integration* (the ability to make use of a variety of external and internal stimuli). Therefore, Shevlin argues, to develop a conscious machine is synonymous with developing a general intelligence system.

The argument that this is a *category mistake* by the principles of this paper, arises primarily because Shevlin calls for a behavior-based definition. That is, he argues that if an artificial system *behaves* in a way that can appear to be generally intelligent, such a system can be seen as being conscious. Now, as will be stated, a strong tenet of work in the CM category is that *mechanism* needs to be examined as a major differentiating factor between categories. As will be suggested, it is the mechanism for the growth of a 'mental' state structure that defines *becoming conscious* which is a central defining factor. In the Shevlin sequence, there is no reference to any mechanism for the development of mental states in general intelligence definitions. Indeed, a generally intelligent machine could be deeply programmed to behave reactively with no internal mentation. To call such a machine conscious, is, therefore, a ***category mistake***.

3. The Conscious Life Category

It not the intention of this chapter to replay the plethora of theories on what it is for a natural creature to be conscious. The intention is primarily to draw attention to the functions that being conscious endows to a living organism, and how this inspires the desire to instill such functions in designed machines. The category mistake occurs when commentators equate conscious machines with living ones. It is not merely a question of the biological/artificial distinction that is at stake (in theory, a mathematical formulation of could represent both), but the error is because the living makeup creates specific drives and needs during the process of *becoming* conscious in the living system which is not the case with a designed machine. Here is a brief example:

A possible (but close to reality) conversation between a scientist and a journalist illustrates this danger. After a lecture on an existing system in the CM category the journalist asks the lecturer, "Yes but your system is not *really* conscious it's just a computer program. When will you make one that is conscious like I am?" The journalist has crossed a category boundary that makes the question absurd. A point of this paper has been to indicate that the products of the CM category are conscious in a machine way, which can be distinguished from the living version but the function of which does for the machine what the natural version does for the living entity. One task for CM researchers is to develop further what "conscious in a machine way" means. Then an appropriate question for the journalist would be "When will one of your machines be able to tell me what it's like to be a conscious machine?"

To summarize, it is suggested here that entities in both CM and CL categories, become conscious through developing experience in their world. It is the difference between their action ability and the nature of their environment that creates major distinctions between the two. That is, the process of developing as a living entity, fundamentally distinguishes CL entities from CM ones.

4. Some history of work in the Conscious Machines Category: Conscious States in Neural State Machines

The early motivation for work in the CM category, came from observing the similarity between the way brain people talk about 'the nervous system' 'mental states' and 'the mind' on one hand (reviewed and summarized in Damasio [2021]), and the way engineers talk of 'neural networks', 'internal states' and 'state structures', on the other. In Aleksander and Morton [1993] it was suggested that being artificially conscious in a machine could be achieved were a robot endowed with a neural state machine with machine states paralleling mental states through learning to depict inner (e.g., stomachache) and external (e.g., a seen butterfly or a heard musical event). This type of effort gave rise in 2001 to a meeting at Cold Spring Harbor among people who were known for contributing to 'the science of consciousness'. While there was little agreement on precise definitions of consciousness between the audience of 21, made up of neuroscientists, philosophers and computer scientists, there was agreement on the following closing proposition. "There is no known law of nature that forbids the existence of subjective feelings in artefacts designed or evolved by humans". This gave rise to CM work in several laboratories. We pursued this work by formalizing the properties of a neural state machine as follows. In Aleksander and Dunmall [2003] it has been argued that the judgement of whether *a machine* is conscious or not should depend on classifying the internal mechanisms that are at play in making the system conscious.

According to the above, such machinery must be capable of learning to sustain the state structures which support being conscious. Recursive neural nets, as mentioned above are appropriate candidates to fit the bill. A categoric knowledge of the supporting mechanism is helpful because it may not be possible to discern consciousness just from the behavior of a system. *How* this behavior is achieved is important and in the context of this paper the discovery of any seemingly conscious external *behavior* in a machine may not infer the presence of consciousness. A knowledge of the mechanism resolves the essential question which is whether the mechanism can support a learning-induced growing depictive state structure. This issue, as often introduced into discussions about being conscious, is confused by the fact that we accept other living creatures as being conscious without referring to their mechanisms or other theories. But this is conferral and not discernment. In the current discussion about categories, this is one form of confusion that one aims to avoid.

In Aleksander and Dunmall [2003] the broad necessary state-based characteristics have been expressed as five intuitions framed as *axioms* relating to an agent A said to be conscious of a sensorially-accessible world S. Central to this is the concept of 'depiction' which is an internal event that reflects the perceptual form of events in S. The axioms are:

1. Depiction: *A can be in internal states that depict parts of S.*
2. Imagination: *A can be in internal states that relate to previously experienced parts of S or fabricate new S-like sensations.*
3. Attention: *A can select which parts of S to depict or what to imagine.*

4. Planning: *A can exercise control over imaginational states in order to plan actions.*

5. Emotion: *A can be in additional affective states that evaluate the consequences of planned actions.*

Working within this axiomatic domain, clearly supports the need for the presence of an inner (i.e., mental) state sustained by a physical medium. This is developed in the next section. Also, a distinction is made between the terms 'the machine which has consciousness' and 'a machine which is conscious'. The first suggests a property possessed by a machine which needs to be placed there by the designer, while the second suggests that the machine employs felt inner states.

Some significant criticisms can be made at this point Neural State Machines are not the only way that internal representations of experience can be created. This is totally agreed but it seemed in the pursuance of this work, that Neural State Machines are such as to easily carry the axioms. The axioms are also not intended to be comprehensive in covering all aspects of being conscious and an invitation is put forward to researchers to extend the list in important areas. The representation of language and its integration with inner and external perception is a fruitful area for further research within the 'axiomatic' framework. A reviewer has kindly pointed out that issues of deliberation and introspection have been left out in the axioms. In the sections that follow we take further the *Depiction* of axiom 1 to indicate that through Iconic learning the inner states of axiom 2 not only depict experience can *be felt* as experience. As a pinprick can be felt through the action of some neurons, experience is what is felt through a vast pattern of pinpricks depicting experience, either sensory or internal (that is, machine-mental). Therefore, the axioms are indeed concerned with introspection and deliberation. In the current paper, a central feature of the sections which follow is to look closely at memory.

Indeed, in Aleksander and Dunmall [2003], it was argued that a physical constraint for a machine to satisfy axiom 1, is that the depiction of S be treated as the conjunction of minimally discernable events. This is another way of saying that in designing conscious machines the neural methodology becomes favored as such events (e.g., a fly landing on a white sheet of paper) can be discerned because it prompts a change of state in a part (even a minimal one) of an internally represented state. The firing of a single neuron in a neural network is the limit of this.

Then, relating to the CM category, it was suggested that a most important feature of a neuromodel is that, through depiction the content of a conscious experience in the machine can be decoded and displayed for human observation". Hence:

The display characteristic. *A machine in the CM category has a designer/user who, to indicate that that the machine is conscious, includes, as part of the design, means for communicating outwards what the machine is conscious of at a point in time in terms of displaying its internal state.*

This, of course, is a current distinguishing factor from the CL category, where access to an entity's internal state is something one cannot achieve with any confidence. Should it become possible someday to display the inner states of a CL entity, this will aid comparisons between CL and CM categories, However, it would go under the heading of something like 'mind reading' only minimally achieved with scanning techniques at the time of writing.

5. The Importance of Growing State Structures: Machines Becoming Conscious

In brief re-stating the Damasio/Engineering similarity:

The State Space Characteristic. *It is convenient and appropriate to assume that the theoretical structure of a Conscious Machine is a state machine[b] (primarily probabilistic, but deterministic in the special case). In other words, the key inner representation of mentation in a CM category entity is a felt state structure, in the state space of the machine.*

One effect of this characteristic is that it positions the mind/body discourse in the framework of the 'material structure/function' character of state machines. In the field of neural automata theory, it is known that the structure/function relationship for neural state machines is not straightforward as it carries two major uncertainties. First, for a given

---

[b]https://en.wikipedia.org/wiki/Finite-state_machine.

neural physical structure (geometry of the net) there can be many different state structures depending on the inner dispositions (for example, *weights*) of the neurons. And conversely, a specific state structure may be supported by many physical structures (the 'state assignment problem'). A key realization in the neural automaton under discussion is that these uncertainties are resolved by the accumulation of experience (learning). It is now shown that the state structure of a CM element can be determined by the accumulation of experience which is represented by a *growth of a meaningful state structure* which encapsulates this experience. An illustration of this process follows below with a suggestion for neural state machine implementation in Sec. 5.

Say that an exploratory robot contains internal machinery that can be in many states. This is a probabilistic system which initially has equal probability of transiting at a given time from one to any other state. An impression of how this could be imagined is shown in Fig. 1. Now, say that the system has a perception w_a of its world. It represents this internally as w_a^i where the latter is a *depiction* of the former (Fig. 2).[c]

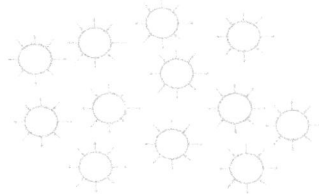

Fig. 1. Part of a state space for a probabilistic state machine before any growth takes place. There is no metric in this space and the circles represent, as yet unlabeled, states from which radiate the equal probability transitions from and to other states.

---

[c]It is important to note the symbol w does not refer to a weight as it might do in other neural network literature. Here it is used to indicate a state machine input that causes an inner state transition (labelling the transition arrow in the state diagram) and for Iconic training (Sec. 4) it is also part of the label of the state (circle)which is reached through this transition.

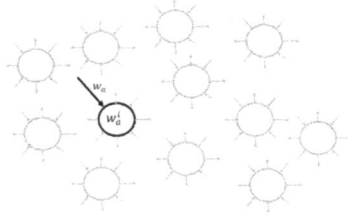

Fig. 2. Perception w_a leads to internal state w_a^i which is the 'depiction' of, w_a that is, there is a close point to point similarity between the two, but they are not identical for practical reasons. It is noted that for the machine, w_a^i is the only knowledge that it has of w_a.

Further exposure and an adaptation ability make this depiction re-entrant at which point it comes into existence in the machinery as belonging to that machinery as a stable internal state and thus becomes an element of the state structure of the machinery. What this means is that even with perception switched off (φ) the machine can remain stable in w_a^i (Fig. 3).

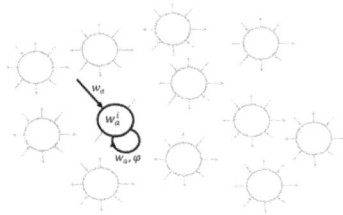

Fig. 3. A stable depictive internal state, sustainable without input is created by switching the input to some unspecified value, φ. Noise due to a blink in vision is an example of this. Note that closed circles represent states, where open, arrowed, circles represent transitions that go back to the originating state. Such states are called stable states and the arrowed ones are called *reentrant* transitions, the reentrance being valid for the inputs for which the reentrant arrow has been labelled.

The CM researcher will assert that, though this property, it is appropriate to say that *the machine has become conscious of* w_a by 'owning' the depiction w_a^i in its state structure. The experimenter, having engineered the presence of the *display characteristic*, can check that even after noisy conditions and no perception (φ), state w_a^i remains stable within the machine or is one of such states, should there be many of them. The statement that *the machine has become conscious of* w_a must be interpreted as a technical statement and, not as the presence of a human property. It merely means that there is a *functional identity* between the machine property in an element of the CM category and that of an element in the CL category. The style *m-conscious* is therefore suggested to be used for the CM category (*the machine has become m-conscious of* w_a).

But, so far, there is no *state structure* only the vertices of some graph that such a structure *will become*. While being m-conscious of w_a^i say that the perception changes to w_b. Then, not only does the system learn to be m-conscious of w_b (by creating the internal stable state w_b^i) but also, that w_a^i is a pre-condition for this event. In state machine theory, one would say that 'a transition has come to exist between states w_a^i and w_b^i mediated by w_b' as part of the growing structure of becoming conscious of sequences of perceptual events. The CM researcher would express this as *the machine is m-conscious that* w_b^i *has followed* w_a^i. This is represented in 'state space' as shown in Fig. 4.

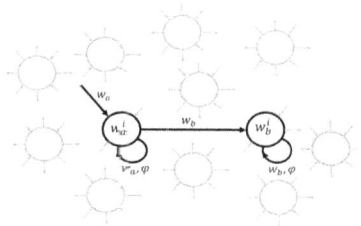

Fig. 4. The machine *becomes* m-conscious of the fact that, when in w_a^i a transition occurs to state w_b^i. The diagram represents the *state structure* created so far according to the way that such structures are represented in automata theory.

In the above, no cause has been attached to the change of perception. Now allow that the machine can apply actions to its world. Say that a set *A* of *n* actions is available to the machine:

$$A = \{a_1, a_2, \dots a_n\}.$$

To begin with, actions are available in random fashion. This assumes that the machine has external actuators that can act on the world. These are activated but in a way that is curtailed by the world. If one stands at a fork in the road the choice of action may only be two, say a_l and a_r (for left and right). If one move is taken by the action mechanism, it is initially sensed as an external event and coupled in state space (arrow) with the change in perception. That is, it becomes associated with the transition to another state (Fig. 5). Again, the learnt transition can be sustained in the absence of perception, (φ). The key point of this discussion is that it shows that *'learning'* in a machine of the CM category is the **growing of the depictive state structure** of an automaton. As time progresses such a state structure represents all that the machine can be m-conscious of with each state being what it is to be m-conscious of at some point in time.

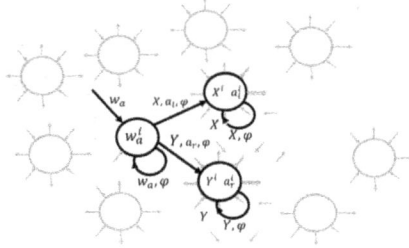

Fig. 5. The machine in state w_a^i is m-conscious of the fact that action a_l leads to world state X and a_r leads to world state Y. (Note, this does not explain *how* the system selects an action to achieve a desired result: this is part of a separate discussion).

Figure 5 also draws attention to the fact that in the φ mode the machine becomes probabilistic (indicated by more than one φ exits from each state, each such exit having a probability) and is in a deliberative mode, capable of exploring the inner state structure. This is the subject of current work.

6. The Conscious Machine Category and State Machines

While it was stated above that the concept of minimal discernable events prompts a neural implementation, it is not implied here that this is mandatory for a machine in the CM category. However, many properties of neural nets contribute to the satisfactory implementation of important features in this category. First there is the *generalization property*. That is, if the machine has learned a transition to state w_a^i under perception w_a there exist other perceptions $w_c{}'$ which will also produce a transition to w_a^i without any further learning. These perceptions are said to be *similar* to w_a. As an example, this kind of property is typically useful when w_a^i is a prototypic depiction, say, of a face that is triggered by many perceptual versions of the input (i.e., various forms of $w_a{}'$).

The property is also associated with the '*stability*' of neural networks: that is, the fact that experience-representing states, such as the re-entrant w_a^i and learnt transitions, become events in the state space that can be rapidly visited under non-perceptual conditions, φ. That is, they have meaningful *attractor* properties.

It now becomes possible to suggest a set of characteristic issues that are definitional of the CM category. While above, discrete state space is implied requiring discrete state equations, state machine formulation does not exclude continuous state-space models as used in control system theory. The key feature is that a state variable be related to a minimal element of conscious mental experience, that is, the output of a neuron fits well with this requirement has indicated earlier.

CM1: Studying the *current mental state* of a system involves a study of the current *state of a state machine.*

CM2: The current state of a machine is said to be m-conscious if it *depicts* a current or previous perception (whether external, e.g., vision or auditory or internal, e.g., pain or nausea).

CM3: *Becoming depictive* of perceptual events is the primary aim of *learning* in the system. An example is 'iconic learning' in neural nets (see Aleksander and Morton [2012], p. 94). Here, internal neurons learn to represent perceptual input states.

CM4: *Learning* is also the depictive creation of transitions between internal depictive states. Such states and transitions form the *state structure* of the machine. This state structure is the repository of accessible experience (the mind in the machine?).

CM5: The state structure can be *deterministic* or *probabilistic.*

CM6: The system can *act* on its world and thus lead to a change in the perceived world. This action is linked in state space to the transition between the states that it causes. Note that perceptions can change with no action. This is simply represented by a null value for the action of that transition.

CM7: The state structure can be traversed internally when there is no perceptual influence (the φ condition) creating a contemplative function.

This can now be stated as a central assertion that characterizes the CM category:

__The CM Assertion__: That of which a conscious machine is m-conscious, exists as a structure of inner states within the machine. This grows to depict the experience of the machine as such experience progresses. Some of this progress is in function of the machine's actions. A display allows a researcher to monitor the development of states.

7. A Neural State Machine Memory Model in the Conscious Machine Category

As an example of the use of neural technology for modelling in the CM category, models of human memory modelling are addressed which, traditionally, are largely based in the AI category. We develop this illustration from the CM category by applying the principles outlined in Sec. 6, above. Primarily, the *CM Assertion* is invoked to argue that conscious features of memory such as the transition between short- and long-term memory and fading of short-term memory, can be studied as state structure effects in a physically realizable neural state machine (NSM).

Specifically, two experiments are pursued: first, the fading of the memory of a single perceptual event and the conditions for fading not to take place with long-term memory replacing the short and, second, factors affecting the capacity of long-term memory in the memory of sequences of perceptual states.

Recognizing earlier important AI category work one notes the models of short- and long-term memory (STM and LTM) due to Atkinson and Shiffin [1968]. This established the notion that short- and long-term memory are *physically distinct* processing functions with data flowing from the one to the other. A rehearsal process takes place which can sustain a memory in STM in preparation for the transition to LTM.

The influence of this type of model can be discovered over a vast sector of memory literature stretching to current material (as surveyed by Plancher and Barrouillet [2020], for example). Even in an important sector of the machine consciousness literature (see LIDA in Baars and Franklin [2009]) many memory functions are seen as being physically distinct and brought into consciousness in a Global Workspace through a competitive process. In the CM state machine model, these distinct memories occur in different areas of *state space* and not physical space and enter consciousness by entry into a state in one of these regions.

It is noted that the work in the current paper is distinguished from interest in Long-Short-Term-Memory (LSTM) as it is the depictive nature of an internal state that is at stake in terms of its permanence or volatility as it affects that of which the machine may be conscious. In LSTM time parameters are construction tools that are used to effect desired aspects of system performance. The thrust of this section of the paper is to investigate how memory duration and permanence arise from the parameters of the net.

7.1. *Memory of single perceptual events*

In the NSM model, the process of creating an internal state that 'remembers' a perceptual event (as in Fig. 3, above) may be annotated as follows. Recall the formulation of the creation of the reentrant state (A1) in Appendix A:

$$(\boldsymbol{w}, \boldsymbol{s}) > net > \boldsymbol{s}'$$

(This reads 'with the net in state s and an input w the net enters state s'.)[d]

For this to be a reentrant state, the following training sequence is required that

$$(\boldsymbol{w}, \boldsymbol{s}) > net > \boldsymbol{s}$$

Then, if the internal state is to *depict* the perceptual input w it is seen in (A2) it becomes necessary that $\boldsymbol{s} = \boldsymbol{w}$. Then for this to become a

---

[d]Bold symbols like $\boldsymbol{\varphi}, \boldsymbol{w}$ are now used to indicate system arrays present in the machine where previously in this paper as φ, w they indicated abstract unimplemented messages.

'memory' of w. However, it was seen in Sec. 3 earlier that for a memory to be created

$$(\varphi, w) > net > w \qquad (5.1)$$

and φ has been described as **'no perceptual input'** which needs definition. For example, it could be a single state such as an all black perception in the visual domain. While this creates re-entrant states that remain stable under the black input φ producing LTM, there is no STM, as an input of φ directly 'forces' the net to be 'stuck' in w. An alternative is to make φ more complex: a *sequence of n noise* patterns present during training. This is named to be an ***n-wink***. A brief discussion on how this might relate to the *'attentional' blink* that appears in the vision literature is included later.

Here it is best to illustrate the idea with the results of some experiments. Without loss of insight and assuming *visual* perceptual inputs, φ is modified to θ^n, which is the symbol for an ***n-wink***. Equation (5.1) now is written

$$(\theta^n, w) > net > w \qquad (5.2)$$

that is: training a state to reenter is repeated n times with a different random pattern at the perceptual input each time. Referring to figure A in appendix A, the way that this repetition is achieved and (5.2) is implemented, implies a retention of w at the 'desired state' input while the ***n-wink*** occurs.

To formulate the process of testing the memory of the system assume that the following reentrant states have been created for single perceptual events $w^1, w^2 \ldots w^m$. According to (5.2). To test the retention of the system for its memory of perceptual event w^j, w^j is entered at the perceptual input and the system is allowed to run until the reentrant internal w^j is reached due to its attractor property (In anthropomorphic language: "until the NSM has seen and become conscious of w^j"). Then an 'Eyes closed' state, is created by switching the perceptual input to a long-lasting arbitrarily chosen random input pattern θ_{ec}, that is, one chosen from the entire set of possible input patters. This is not necessarily one in the θ^n sequence. The NSM is then "in a state of remembering w^j with eyes closed". It is shown in subsequent pages that this could be a case of STM with the inner w^j morphing into some meaningless state sequence or LTM where the inner state w^j persists indefinitely, perhaps embedded in some noise.

An experiment of this behavior is described and illustrated. The hypothesis which is tested is that the distinction between STM and LTM is a function of n in Eq. (5.2). On the basis of the following empirical argument.

Fig. 6. A pseudo-realistic scene with a movable window. It has 8 colors shown in grey scale for publication. The movable window (shown over the 'mouse' in the bottom left quarter of the scene) provides an input of 98×98 8-colour (3-bit) perceptual inputs to an NSM.

The $n = 0$ case. Let the system be in reentrant depictive state w^{ji} triggered by perceptual input w^j (according to (A2) in Appendix A). As the input is switched to 'eyes closed', θ_{ec}, the lack of match between this and the previous net input w^j (at a) generates noise to the next w^{ji} which feeds back into b creating more noise, producing a state run-away to some region of state space in which the depiction is lost, that is, forgotten.

The $n \neq 0$ case. As n is increased, the neurons build up a repertoire of random input states that increase the likelihood of a match with θ_{ec} creating, in turn, a reduction of noise in w^{ji}. There is a point at which the noise is insufficient to sustain the runaway situation and w^{ji} becomes the LTM of stimulus w^j.

Igor Aleksander

To illustrate this experimentally, Fig. 6 shows the input arrangement of a pseudo-realistic scene (i.e., with a variability that approaches a photographed scene) with a movable window that provides input to an NSM.

Eye open   Eye closed

State behavior for n = 0						
State behavior for n = 4						
State behavior for n = 8						

t = 1   t = 2   t = 3   t = 4   t = 5   t = 6   t >> 7

Fig. 7. It is shown that the wink-length *n* determines the length of a short-term memory for low values and, for values beyond a threshold (8 in this case), allows the reentrant state to remain stable to form a long-term memory.

The experiment proceeds by moving the window to frame each of the 6 heads of animals as inputs and creating a reentrant internal depictive state for each of them and each considered value of *n*. Testing then takes place (again for each animal and each value of *n*). A set of results is seen in Fig. 7. Each of these starts with a shot of the animal input and the following θ_{ec} input (eyes-closed random input.). Then follow the resulting internal states. It is evident that for *n* = 0, the noise levels increase as the net transits to new states ending in in a set of states related to a mix of all trained reentrant states representing the 'forgotten' situation. Increasing n to 4 reduces the initial noise levels, as expected and the character of image character is sustained for longer, again as expected. Then, again as expected, the noise level generated for a larger n (8 in this case) the noise generated is not sufficient to sustain the runaway and the image is long-term remembered.

In colloquial discussion, it can be said that the length of a perceptual wink during the effort of creating an internal, conscious memory of a perceptual input, determines both a level of Short-Term Memory or the cut-in of Long-Term Memory. **These are different phases of operation of <u>the same neural structure</u> rather than seeking STM and LTM <u>in different</u> structures**.

7.2. *Memory of <u>sequences</u> of perceptual events*

Here, the notion of phases that distinguish between STM and LTM is shown to depend on n for attempts to memorize *sequences* of images of the animals in Fig. 6. The particular sequence selected for this experiment seen in Fig. 6 as the clockwise sequence of animals {yellow mouse — blue mouse — red butterfly — blue butterfly — yellow cat — red cat}. This is written in sequence as $\{w^1, w^2 \dots w^x, w^{x+1}, \dots w^6\}$.

Adopting (A2), each training step links the memory of a perceptual event to the next one as

$$(w^{x+1} \, w^x) > net > w^{x+1}.$$

Which reads: the current perception as the inner state, combined with the next perception leads to an inner state depictively equivalent to the current perception. A rehearsal step is taken by linking the last state (w^6) to the first one (w^1) allowing the sequence to exist in the NSM state space as a self-sustaining entity. The memory of such learning is tested by using a trigger perception (yellow mouse in this case) as an input event, letting this become the internal state then letting the input perception become the 'eye closed' event, θ_{ec}. The resulting state performance is shown in Fig. 8(a). It is seen in Figs. 8(b) and 8(c) that the wink again can create STM and LTM depending on the value of n.

(a) State behavior after 6-animal training, n = 0

(b) State behavior after 6-animal training, n = 4

(c) State behavior after 6-animal training, n = 8

t = 1   t = 2   t = 3   t = 4   t = 5   t = 6   t = 7

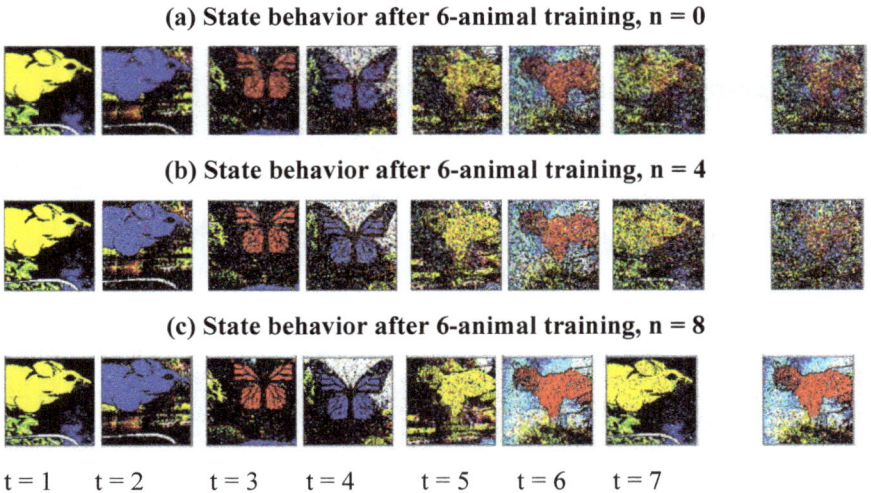

Fig. 8. (a) For *n = 0* it is seen that the sequence creation training process works, but the noise runaway is considerable and the sequence is forgotten by the return to the first pattern and collapsed into a few 'mixed' states of which one is shown on the right of the sequence. (b) For *n = 4* a slight noise improvement occurs, while indicating STM behavior. It also collapses into a few 'mixed' states that indicate a forgotten STM. (c) For *n = 8* shows a retention of the sequence due to *n* being sufficiently large to produce noise below a threshold value to sustain the remembered sequence.

7.3. *Comments on the Memory Model experiments within the CM category*

The intention of the above memory experiments is to demonstrate the appropriateness of the ***Display Characteristic*** of the CM category. The user of the CM is capable of noting from the display of state behavior how introduction of noise in the learning experience of the NSM leads to a slide from STM to LTM in this system as a phase change of activity in one net rather than a transmission between separate LTM and STM systems. In general, knowing 'what a machine is conscious of' is a central feature of machines in the CM category. For example, in the design of autonomous vehicles that embrace CM methodologies keeping a memory of mental states and resulting actions is central in determining whether the machine reacts appropriately to that of which 'it is conscious'.

8. A Look at Work in the Conscious Machine Category

The topic of conscious machines has a history of approximately 30 years. Here one looks back on some of this work to examine how this earlier work fits in with the CM category. An early foundation was laid by Bernard Baars [1988] featuring Global Workspace Theory (GWT). While not adopting the language of automata theory, the Global Workspace is, indeed, the holder of a current mental state on which depends the transition to one of a selection of other states determined by a competitive process influenced by external input. In contrast to the structure given here, symbolic methods were used and memory processes (Semantic, Procedural etc.) are described by Stan Franklin and his team in the context of LIDA, a proposed implementation based on GWT (Baars and Franklin [2009]). On the neural side, Jean-Pierre Changeux and Stanislas Dehaene, used showed that GWT could have a neuronal structure. This gave early examples of the interplay between the CM and CL categories (for example, neuronal processes responsible for verbal reports. See, Dehaene, Sergent, and Changeux [2003]). The work of Seth [2008] on measurements of consciousness is of much interest.

Another early CM contribution is the neural architecture due to Pentti Haikonen [2003]. This is very close to the generic neural automaton described above, with the major difference that it consists of pre-defined interacting automata which deal separately with different perceptual modalities. While mechanisms of neural learning are discussed, how state structures grow, is not considered.

Within the CM category, important work has been done on physical robots. Antonio Chella [2007] has developed the Cicero museum guide robot and has shown that taking an 'externalist' standpoint (the external content of the perceptual loop is included in models of being conscious, see Manzotti [2006]), leads to practical machinery. This is relevant to the definition of the CM category under CM6 in Sec. 4 earlier, which links growth of experience to action and exploration. Owen Holland *et al.* [2007] discuss CRONOS, the human-like muscle-and-skeleton model designed to study embodiment in conscious machines. They draw attention to the importance of having a model of world and self as a simulation internal to the robot. In this paper it is proposed that this internal model be the growing probabilistic state structure of a neural automaton.

9. Conclusion and a Glance at the Future

It has been asserted that conscious machines constitute a specific category (CM) in which 'being conscious' stands in relation to the designed nature of the machine in a way analogous to the way that 'being conscious' in a living organism stands in relation to the biological nature of a living conscious organism, that is, one that belongs to the (CL) category. Importantly, category CL is *totally distinct* from CM except for two factors. The first is that CM can act as an analogical explanatory medium which helps with the formulation of theories about CL, and the second is the technological step that CL may inspire the design of CM machinery (robots and the like) through advantages that being conscious brings. Both these categories stand aside from the AI category which is generally driven by the ingenuity of a system designer rather than the process of *becoming* conscious through growth of structure in state space.

Concentrating on CM, possible future work becomes evident. As the use of deep learning in substantial neural structures gives good performance in current AI (given sufficiently large training data), in a CM system a similar procedure can be applied. Given a substantial neural automaton and a simulated world of enough complexity (possibly virtual), the neural automaton can autonomously become conscious of the world through the process of state structure growing described above. The simulated worlds could contain both external and internal, objects, e.g., intentions, pains, likes and dislikes.

This opens the opportunity for new future studies in the CM category that benefit from machines that become conscious, and investigate:

> ➤ understanding language (the explored world would contain language),
> ➤ animating implanted medical nano-robots,
> ➤ making autonomous vehicles conscious rather than being pattern recognizers,

- ➤ becoming conscious of abstractions,
- ➤ becoming conscious of the characteristics of living objects in the world and communicating with them,
- ➤ etc.

But above all, a clear but collaborative distinction between the three categories defined in the paper can be explored to create a novel multi-category computational paradigm.

Appendix A: Neural State Machines as Workhorses for the Conscious Machine Category

The function of the learning neural state machine is visited in Fig. A, below. This shows the essential components and connections of the machine as it has featured in previous publications on 'weightless' neural systems.

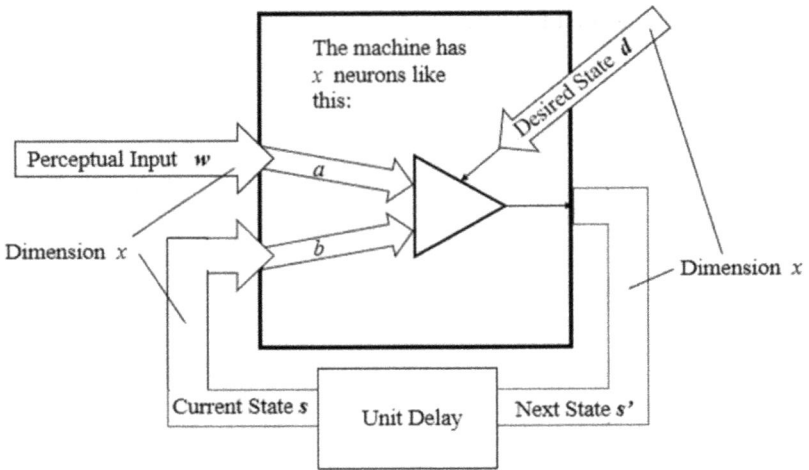

Fig. A. Basic structure of a Neural State machine.

The machine is composed of x "simplified" neurons which through their output connections form the internal state s' of dimension x. A simplified neuron delivers a logical form of learning, by storing inputs as indices to desired outputs met during learning phases. When 'used' and not learning, the neuron finds *a near neighbor* (Hamming Distance) to the current input, among the stored learned inputs, and then outputs the stored desired output. The amount of nearness is a machine variable open to the experimenter to choose.

In this case, each simplified neuron receives a inputs sampled from an overall input to the machine and b inputs sampled from the internal state of the machine which consists of the outputs of the neurons, thus creating the recursion via a unit delay which separates the current state from the next state. Being a discrete time machine, the current state becomes the next state at the arrival of a timing pulse. Inputs and outputs are discrete and carry one of m possible messages. Learning occurs at a given moment through the neuron being given a desired output (single data point) sampled from an *s-dimensioned* desired next state input. This desired output is associated with and indexed by the *a, b* input. Generalization occurs as the 'nearest neighbor' of an unknown a, b input is found and the associated d message is generated. This allows a generalization departure from trained inputs to similar ones. There is also the possibility that in the activation of a nearest neighbor several are found with conflicting stored d messages. In this case the neuron outputs a randomly selected output from m. The overall learning function of the neural state machine is generally

$$(w, s) > net > s' \qquad (A1)$$

w reads: the perceptual state,
s and s' read: current internal state and 'next' desired internal state respectively,
$> net >$ reads "the mapping function mediated by learning in the neural network".

For re-entrance, $s' = s$,
Depictive learning in the net is represented by

$$(w, s) \overset{net}{\Longrightarrow} s' = d(s') = w^i \text{ (The internal depiction of } \mathbf{w}) \qquad (A2)$$

This indicates that equating the desired-state input to the next occurring perceptual state, gives the machine the *depictive* character discussed earlier which is central to the categorization of the machine in the CM category.

References

Ryle, G. [1949] The Concept of Mind (Hutchinson University Library, London).
Shevlin, H. [2019] "To build conscious machines, focus on general intelligence: A framework for the assessment of consciousness in biological and artificial systems," in *Towards Conscious AI Systems Symposium, (AAAI SSS-19, Palo Alto, CA)*.
Damasio, A. [2021] Feeling and Knowing: Making Minds Conscious (Pantheon Books, New York).

Aleksander, I. and Morton, H. [1993] Neurons and Symbols: The Stuff that Mind is Made Of. (Chapman and Hall, London).

Aleksander, I. and Dunmall, B. [2003] Axioms and tests for the presence of minimal consciousness in agents, J. of Consciousness Studies 10(4–5), 7–18.

Atkinson, R.C., Shiffrin, R.M. [1968]. "Human memory: A proposed system and its control processes". In Spence, K.W., Spence, J.T. (eds.). *The psychology of learning and motivation.* 2. New York: Academic Press. pp. 89–195.

Plancher, G., Barrouillet, P. [2020] On some of the main criticisms of the modal model: Reappraisal from a TBRS perspective. *Mem Cogn* **48**, 455–468 (2020). https://doi.org/10.3758/s13421-019-00982-w.

Aleksander, I. and Morton, H. [2012] Aristotle's Laptop: The Discovery of Our Informational Minds (World Scientific Press, Singapore).

Sternberg, R. J. [1977]: Intelligence, information processing, and analogical reasoning: The componential analysis of human abilities (Erlbaum, Hillsdale, NJ).

Baars, B. J. [1988] A Cognitive Theory of Consciousness (Cambridge University Press, Cambridge).

Baars, B. J. and Franklin, S. [2009] Consciousness is computational: The LIDA model of Global Workspace Theory. International Journal of Machine Consciousness, 1(1), 23–32.

Dehaene, S., Sergent, C., and Changeux, J.-P. [2003] A neural network model linking subjective reports and objective physiological data during conscious perception. Proceedings of the National Academy of Sciences 100(14):8520-5.

Haikonen, P. O. [2003] The Cognitive Approach to Conscious Machines (Imprint Academic, Exeter).

Chella, A. [2007] Towards robot conscious perception, in A. Chella & R. Manzotti (eds.), *Artificial Consciousness* (Imprint Academic, Exeter), pp. 124–140.

Manzotti, R. [2007] From artificial intelligence to artificial consciousness, in A. Chella & R. Manzotti (eds.), *Artificial Consciousness* (Imprint Academic, Exeter), pp. 174–190.

Owen, H., Knight, R. and Newcombe, R. [2007] in A. Chella & R. Manzotti (eds.), *Artificial Consciousness* (Imprint Academic, Exeter), pp. 174–190.

Seth, A. K. *et al.* [2008] Measuring consciousness: relating behavioral and neurophysiological approaches, Trends in Cognitive Science 12(8), 314–321.

Haikonen, P. O. [2003] The Cognitive Approach to Conscious Machines (Imprint Academic, Exeter).

Manzotti, R. [2007] From artificial intelligence to artificial consciousness, in A. Chella & R. Manzotti (eds.), *Artificial Consciousness* (Imprint Academic, Exeter), pp. 174–190.

Chapter 3

A Constructive Explanation of Consciousness and its Implementation

Pei Wang

Temple University Department of Computer and Information Sciences, College of Science & Technology, Temple University
pei.wang@temple.edu

NARS is a general-purpose AI system with unified reasoning and learning abilities. This model of intelligence also provides a constructive explanation of consciousness by having partial awareness and control of its internal processes, while being unconscious of the other processes that run automatically. This model acknowledges the distinction of the phenomenal and the functional facets of consciousness but treats them as the first-person and the third-person perspectives, respectively, of the same underlying process.

Keywords: adaptation, insufficient knowledge and resources, reasoning, learning, self-awareness, self-control, NARS

In recent years, consciousness has become a hot topic in several fields, and opinions from various perspectives have been proposed [Blackmore (2004); Gamez (2008); Chella and Manzotti (2012)].

Like many basic concepts about thinking and mental processes, "consciousness" has no widely accepted definition, and in different fields, the focuses of research are not the same. This article will not address all approaches, however, it will describe our approach toward this topic. This line of work has been introduced in our previous publications, such as [Wang *et al.* (2017, 2018)], and this writing, as an extended and revised version of [Wang (2020)], will focus on new considerations and progress in our project that are not covered in the other writings.

I will start by explicitly stating my epistemological presumptions and their implications. Consciousness will be introduced initially as a cognitive function that can be realized in a computer system. I will then briefly summarize the design of our model, NARS (which has been specified in [Wang

(2006, 2013)] and explained in our previous publications), with a focus on the distinction between *conscious* and *unconscious* processes. Finally, I will address the phenomenal aspect of consciousness, as well as the related theoretical issues in several disciplines.

1. Theoretical Position on Descriptions

Like it or not, *consciousness* is fundamentally a philosophical issue in epistemology, as it is about the nature of our knowledge or *descriptions* of ourselves, other minds, and the outside world. As this article is not a purely philosophical essay, here I will simply state my position without a systematic comparison with the existing (huge) literature on this topic. Some comparisons with the most relevant opinions will be made later in the article.

1.1. *Description: subjective vs. objective*

To summarize my position on this topic in one sentence, it is *"There is no describer-independent description of any entity or event."*

I acknowledge the existence of the outside world as independent of any observer. However, any *description* of an entity or event (not to mention the world as a whole) inevitably depends on the describer (call it "agent" or whatever) making or holding the description as belief or knowledge. This dependency comes from several sources:

- The sensorimotor interface between the agent and its environment decides the primes (the atomic sensations and actions) the agent can directly experience or take.
- The available concepts within the agent decide the patterns or schemata used in organizing the experience.
- The communication language used by the agent in social interactions decides the vocabulary that consists of the indirect or social experience of the agent.
- Since an agent is always limited by available computational resources, every description is selective in content, influenced by the agent's motivational complex, attention allocation, emotional status, etc. at the moment.
- As the environment changes constantly and unforeseeably, any summary of the past experience of any agent may be confronted by new experiences.

Because of these reasons, there is no way to "describe the world (or part of it) as it is." Instead, every description is made by an agent at a certain moment from a specific perspective. Here "description" also includes "belief", "knowledge," "theory" and similar notions.

Therefore, in its preliminary form, a piece of knowledge K is *subjective* by nature and should be considered as expressing "I feel K," "I believe K," etc., and is from the viewpoint of an agent, as it comes from the system's experience. Here the word "experience" is used in its basic sense, meaning a record of the agent's interaction with the environment in a comprehensible form to the system. Similarly, the meaning of a concept is also subjective, as it indicates the role played by a specific ingredient or fraction in the agent's experience.

As soon as an agent begins to communicate with other agents, involved concepts start getting a shared, or *inter-subjective* aspect, otherwise no mutual understanding or cooperation is possible. Even so, the subjective aspect is still there. Gradually, some subjective beliefs of the form "I believe K" may become "Everyone believes K" if it is confirmed in the communication process by the others. which can be simplified into "K," and even be interpreted as an "objective fact" that does not depend on any agent. Since people have different experiences, they rarely agree on anything completely, so this type of objectivity is always a matter of degree. Furthermore, even this objectivity still depends on the community of agents participating in the communication. For example, in different cultures what is considered as "fact" is not exactly the same.

Denying the existence of objective fact not only contradicts some philosophical theories but also challenges the widely accepted opinion that some of our knowledge, especially a "scientific theory," describes the world *truthfully*, rather than shows someone's personal opinion or preference. Of course, my above position does not mean that all theories are equally good (as none of them is true). For a certain situation with problems to be solved, each candidate theory has a degree of applicability, jointly determined by its *correctness* (evidential support), *concreteness* (instructiveness), and *compactness* (simplicity), as argued in [Wang (2012b)]. Even the best theory is still not objective descriptions of the domain.

Though science was traditionally associated with objective truth, the dependency of description on describers or observers has been revealed by more and more evidence from fields including cognitive science, philosophy of science, quantum mechanics, and so on. This opinion has not become the common sense of the general public, mainly because the traditional

opinion works well enough in our daily life for most purposes. Even so, for
the current discussion, such a treatment is no longer acceptable. On the
contrary, it is the root of many problems, especially that of consciousness.

The study of artificial intelligence raised the possibility of a type of
agent or describer that is very different from human beings, though still
can "describe" the environment in the broad sense of the word. Given its
electromechanical nature and its relation with us, their sensorimotor experience and social experience will be very different from ours, though corre-
spondence and translatability still exist to a certain extent. Consequently,
our consensus with them will be even less than that among ourselves. It
means that some of our "objective facts" will have to be considered as "human opinion," and their different descriptions of objects and events cannot
be taken as wrong, simply because they are different from ours.

1.2. *Relations among descriptions*

If the same subject matter gets different descriptions, there are various
situations that should be distinguished, as they correspond to different
relations among the descriptions.

When two descriptions mainly include the same concepts and conceptual relations but assign different truth values to those relations, they are
conflicting and *incompatible* with each other. If they come from different
experiences, the conflict may be resolved by pooling the evidence and summarizing the descriptions into a single one with a truth value corresponding
to the pooled evidence, or by selecting the description that is applicable to
the current situation and ignoring the other as irrelevant.

When two descriptions mainly include disjoint concepts, they may provide complementary summaries of the subject matter, though in certain
situations they also compete as better ways to perceive it. In this situation, the important question is not which of them is more *correct*, but which
is more *suitable* to serve the current need.

When the descriptions are composed of a large number of related sentences, they can be considered as *theories* about the subject matter. In
most domains, there are usually multiple theories that not only describe
the situation differently but also propose different predictions and recom-
mendations for actions. Usually, each theory has its strengths and weak-
nesses, which restrict its applicable situations. Contrary to a naive belief,
competing theories cannot be combined by merging the "correct parts" of
each into a new theory, because the concepts in them often correspond to
incompatible ways of summarizing the phenomena.

For the current discussion, the most relevant case is when there are multiple theories at different levels of abstraction or generalization. The human brain is often described as a neural network, where the basic unit of analysis is a neuron, with its internal structure (soma, dendrites, axon, and input-output mapping) and external connectivity (with other neurons). However, this is not the only level to describe what happens in a human brain. It is clearly possible to describe a brain as consisting of molecules, atoms, and even quanta. On the other hand, it is also possible to describe the human *mind* (rather than *brain*) using the language of psychology (and philosophy) by talking about concepts and their processing.

Though the above levels of description are well-known and uncontroversial, some misconceptions about their relation are widespread and even taken as self-evident. The most eminent one is reductionism, which takes a lower level as more fundamental or real, and a theory at a higher level can, at least in principle, be completely translated into a low-level theory, but not vice versa. In cognitive science, this school of ideas includes the belief that all psychological phenomena can and should be explained using neurological phenomena and their relations because the former are *caused* by the latter. Some researchers believe that eventually every theory can be absorbed into a "Theory of Everything" in the language of physics, to which all other theories are merely approximations or special cases.

This belief is not completely groundless, as many psychological concepts and phenomena, including those related to consciousness, indeed have neural-level explanations [Baars (1988); Dehaene (2014)], and similarly, physics provides explanations for the concepts in some other disciplines. However, as a general conclusion, reductionism assumes a "true description" of everything and therefore contradicts the epistemological postulate I proposed previously that there is no objective description of any entity or event.

When the same subject matter is described at different levels, a low-level description usually contains more details within the subject, while a high-level description usually contains richer relations between it and its context. Consequently, in general, neither can be fully replaced by the other, though for a specific purpose, one of them may work better. This is just like observing objects through lenses with different magnifications. Even for the same object, each lens provides a different vision, with its scope and granularity. Each of these visions is different from the others and needs to be described using its corresponding vocabulary. The important point here is that none of the visions is "truer" than the others, or is "fundamental" so that the others can be derived from it.

One factor often omitted by the supporters of reductionism is the cognitive capability of the creator or user of a theory. A low-level theory does provide more details, but at the same time has a more restricted scope. It is also unsuitable to describe large-scale patterns and regularity that are visible only when many details are omitted. It is just like you cannot expect a tourist to use a 1:1 city map, even though it contains much more information than the maps available at visitor centers. I conjecture that there is a rough upper bound on the number of concepts a theory can have, beyond it the theory will be too hard to comprehend and use for a normal human mind. It is similar to the case that even though every software is eventually coded by a long binary string, we cannot, even in principle, discuss software design by talking about which bit should be put at which position in the string. For this reason, there cannot be a Theory of Everything, as there are needs for theories at different levels of description and with different focuses, and these theories are organized around disjoint central concepts and cannot be fully reduced into one another.

Another common misunderstanding is to consider a high-level phenomenon as *caused by* certain low-level phenomena. As in the lenses metaphor where the visions from different lenses do not cause each other, theories at different levels may describe the same event differently, but since they are (parallel) descriptions of the same event, there is no causal relation involved — though causality has different definitions, it is nevertheless about the relations between *separate* events. People indeed often accept a low-level explanation to a high-level concept (just like to "zoom-in" at a certain point in the lenses system), but such explanations are not *causal*. Furthermore, this "zooming-in" practice does not mean that a low-level theory is superior to a high-level one by being "more informative" or something like that because the opposite "zooming-out" practice also helps in understanding an entity or event by putting it in a wider context or larger picture.

Similar opinions like my above one are often proposed by claiming the high-level phenomena are "emergent" from the low-level, so cannot be explained in a reductionist way. Though I agree with the conclusion to an extent, such a description still implicitly takes the low level as more fundamental, on which the high level is based, though with its special abstractions or other constructions. Such an explanation is acceptable for certain purposes but has its troubles. Take physics as an example. Even though it is often assumed as "at the lowest level" in the hierarchy of scientific theories, many of its concepts, such as quantum, quark, string, etc., are highly abstract and are constructed from human experience through many mental operations. To claim that all other scientific concepts must be explained in terms of them sounds preposterous.

In summary, theories at different levels of descriptions are surely related to each other in content, as a high-level theory is usually a generalization of a low-level one. However, there is neither isomorphism between their concepts, nor causal relations between their phenomena. In general, each level may have its unique values and usages that cannot be obtained at another level, either below or above it.

1.3. *Self-description*

According to the previous analysis, every description and theory is from the viewpoint of some agent and at a certain level of abstraction. When this conclusion is applied to the self-description of an agent, there is something special. As explained above, even though every description is intrinsically subjective, it can become relatively objective (i.e., inter-subjective) to an extent when shared by a group of agents who have similar experiences with respect to the object of the description.

However, it cannot be the case when the object is the agent itself. In this situation, the self-description is obtained via a different perceiving process compared to a description made by others on the same event. When an event happening in agent X is described by agent Y and Z via observation, as well as by X itself via introspection, the corresponding descriptions D_Y, D_Z, and D_X are all different from each other, though D_Y and D_Z will be much more similar to each other than to D_X in several aspects:

- They are obtained via different sensors.
- They go through different perceiving processes.
- They are categorized and expressed differently.
- They trigger different associations and responses.

Consequently, D_X is often considered as *private* and *subjective* (as the processing mechanism is not shared with anyone else), while D_Y and D_Z are *public* and *objective* (as the processing mechanism is largely shared with other agents). This distinction is basically the same as that between the so-called "first-person perspective" and "third-person perspective." Though there are correlations between the two, their differences are fundamental. There is no way to feel "what is it like to be X" without being X, as argued by [Nagel (1974)].

This distinction is at the core of the puzzle of consciousness, which is closely associated with introspection. I acknowledge this distinction, though disagree with some of its common interpretations:

- It does not mean that consciousness cannot be studied by science, as its private and subjective nature conflicts with the pursuit of science being public and objective. I think that though a conclusion must be public and objective to be considered as a part of science, it does not mean that the *phenomenon* described by the conclusion cannot be private and subjective.
- It does not mean that an explanation of consciousness must reduce a first-person perspective into a third-person perspective. According to the above analysis, it is neither possible nor necessary, because the same process may have different descriptions from different perspectives, and none of them is more fundamental.
- It does not mean that it is impossible for AI systems to become conscious, because an artifact cannot have anything private and subjective by nature. I think this feature just prevents us from directly feeling what it is like to be an AI, but it does not mean that an AI also cannot feel what it is like to be an AI, as it is exactly what it "feels," in the sense of receiving and processing a certain type of signal to obtain some effects. We indeed cannot decide whether an AI is conscious according to a phenomenal standard, but have to explore the other aspects of consciousness. However, we cannot even use such a standard to decide whether another human being is conscious, as we cannot feel exactly what it is like to be that person. Since in this case, we accept indirect evidence (provided by sympathy, analogy, and so on), the same should be acceptable for AI, though the evidence may be collected in other ways.

Therefore, I agree that consciousness has a *phenomenal* aspect, which can only be felt or experienced from a first-person perspective, so cannot be used to evaluate the existence or complexity of the consciousness of someone else. However, I do not consider it the only facet of consciousness. The other facet, call it *access* or *functional* consciousness, can be evaluated scientifically. I consider the two facets as unified as two sides of the same coin, so an agent cannot have one without the other. For an agent, no matter whether it is a human or an artifact, its knowledge about itself can provide cognitive functions in its adaptation process, and makes important differences in its behaviors. Consequently, we can check whether an agent has the functional facet of consciousness, and use it to infer the existence of the phenomenal facet.

2. Consciousness in OpenNARS

In this section, I introduce the functional aspects of consciousness in the AI system NARS (Non-Axiomatic Reasoning System), as being implemented in the open-source software. Since NARS is a complicated project and it has been described in many publications including two books [Wang (2006, 2013)], here I only summarize its features that are most relevant to this topic.

2.1. *Objective and approach*

In my opinion, intelligence should be understood as *the capability of adaptation with insufficient knowledge and resources* [Wang (2019a)], rather than as specific problem-solving skills. Accordingly, NARS is designed under the *Assumption of Insufficient Knowledge and Resources* (hereafter referred to as *AIKR*) [Wang (2011)], meaning that the system manages its finite processing power while being open to novel tasks, and responsive to time pressure. The system is adaptive in the sense that it depends on its past experience to deal with the current situation and to allocate its resources among competing demands in a context-sensitive manner.

To realize this capability in a domain-independent way, or being an *artificial general intelligence* (AGI) [Wang and Goertzel (2007)], NARS is built in the framework of a *reasoning system*. The system carries out all cognitive processes as *reasoning* (also known as *inference*), and each of them consists of inference steps following a formal logic, so is justifiable and explainable, since the same logic also captures the principles of human thinking, too.

The existing logic systems in the related fields are mostly incompatible with AIKR, and usually do not take adaptation into consideration, as analyzed in [Wang (2019b)]. Consequently, the design of NARS includes a novel logic, Non-Axiomatic Logic (NAL), which has been created especially for adaptive reasoning under AIKR.

NAL is a *term logic* in the tradition of [Aristotle (1989)]. In NAL, a *term* refers to a concept within the system (rather than denoting an object or event outside the system) and represents an identifiable ingredient in the system's experience. A *compound term* is formed by a number of component terms in one of the predetermined logical structures (similar to the operations in set theory and Boolean algebra) and represents a useful pattern in the experience. A *statement* is a special type of compound term and represents a certain form of substitutability between two terms, as well as between the two concepts they identified, respectively.

NAL is treated according to an experience-grounded semantics [Wang (2005)]. According to it, the meaning of a term (or concept) is determined by its experienced relations with other terms (or concepts), and the truth-value of a statement measures its evidential support, also according to the system's experience. Consequently, meaning and truth are experience-dependent and subjective, though there are objective (inter-subjective) factors coming from the shared experience with other systems. In this aspect, NARS is fundamentally different from the traditional "Symbolic AI" where a symbol needs to be interpreted as referring to an outside entity to become meaningful [Newell and Simon (1976)].

Since a truth value no longer indicates the extent for a statement to agree with an objective fact, but the extent for it to agree with the available evidence, non-deductive inference (such as induction, abduction, analogy, etc.) is justified in the same way as a deduction, that is, an inference rule is valid, or truth-preserving, in the sense that the truth-value of its conclusion correctly measures the evidence provided by its premises.

The inference rules of NAL not only derive new statements among the existing terms but also construct new terms and concepts to summarize the system's experience in more efficient forms. Consequently, learning and reasoning are carried out by the same process. When focusing on the individual steps, what we see can be naturally considered as "reasoning," because each step follows a formal rule that derives a conclusion from a few premises; When focusing on the processes composed by these steps, what we see can be naturally considered as "learning," because the content of memory is modified according to the system's experience, representing an effort of the system to adapt to the environment.

Similar to logic programming [Kowalski (1979)], in NARS a statement can correspond to an *event* with a temporal attribute, an *operation* executable by the system itself, or a *goal* to be achieved by executing proper operations. Therefore, the representation language of NARS, *Narsese*, can uniformly represent declarative, episodic, and procedural knowledge in a conceptual network, where each vertex corresponds to a concept named by a term, and each edge corresponds to a conceptual relation. This network is dynamically formed and reformed by the system's experience, as its topological structure and the attributes of the vertices and edges can be changed at run-time. Various types of cognitive processes, including planning, perceiving, acting, predicting, explaining, etc., are all carried out by reasoning. For the details of the inference rules, see [Wang (2013)].

NARS handles concurrent inference tasks by time-sharing, though the mechanism is more complicated than that in an operating system. Under the assumption of insufficient knowledge and resources, NARS needs to handle tasks for which no algorithm is available, as well as to respect their associated response time requirements. By definition, in such a situation NARS cannot achieve every task according to a predetermined standard, nor can it search a solution space to find the best. Instead, the system tries to do its best on all tasks (with a priority distribution) according to its experience. The system's computational (time and space) resources are dynamically allocated among the tasks and beliefs, biased by their priority values. As the situation changes, the priority values are adjusted at run-time, so the system's response to a task depends on history and context, rather than only on the system's design.

2.2. *System architecture and working process*

OpenNARS is an open-source implementation of NARS that has been under development for more than a decade.[1] The current architecture of the system is shown in Fig. 1.

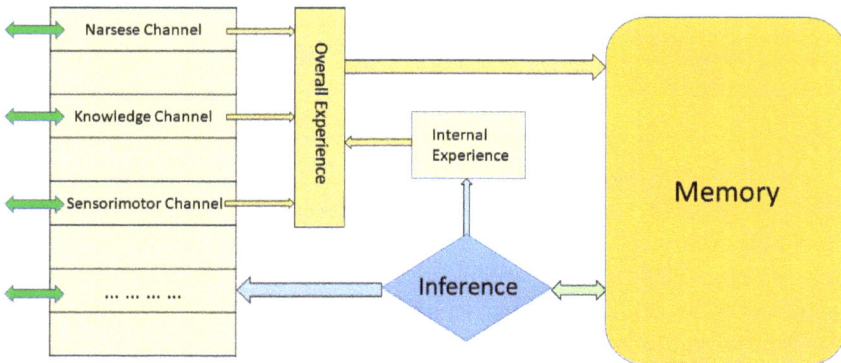

Fig. 1.   Architecture of OpenNARS.

The system interacts with its outside environment via an input/output interface composed of multiple types of channels:

---

[1] URL: www.opennars.org.

Narsese Channel. Each channel of this type connects OpenNARS to a user or another computer system. Each input is a task in Narsese and can be a *judgment* to be digested, a *goal* to be achieved, or a *question* to be answered. In the other direction, normally the same types of tasks can be sent out to the other party of the communication, triggered by operations issues by the inference engine.

Knowledge Channel. Each channel of this type connects the system to a knowledge source, such as a database, knowledge base, ontology server, etc. Such a channel converts Narsese tasks to and from the format used by the other system. The typical working process in such a channel is for OpenNARS to issue a request according to a question to be answered and to translate and evaluate the response obtained from the knowledge source.

Sensorimotor Channel. Each channel of this type connects OpenNARS to a specific type of hardware or software, which serves as a sensor and/or an actuator of the system. OpenNARS operations are converted into commands sent to the device, then the feedback is converted into input tasks of OpenNARS. As discussed in [Wang and Hammer (2018)], perception in NARS is an operation-driven multi-level generalization process, with results at various levels of description.

The channels are storage and processing units that can independently carry out certain common functions:

- New input items are accepted at any moment, as far as they are in the format recognized by the channel. Each item will be converted into Narsese tasks or terms with an initial *priority* according to its features, including the activation level of the corresponding concept in the system.
- Each channel has a fixed capability, and when there are more items, the ones with low priority will be removed. An item will be removed after a certain period of time as well.
- Each channel carries out certain inferences spontaneously on selected terms to form compound terms and tasks. The selection is biased by the priority distribution in the channel. The types of inference include temporal composition, by which a series of events can be perceived as a whole. For channels with multiple sensors, spatial composition also happens, which creates compound terms for certain spatial arrangements of concurrent events.

- In sensorimotor channels, simple sensorimotor contingencies are formed as temporal implications among selected events, which will be used to predict or to intervene in the occurrences of certain events. The truth value of implication statements is revised according to whether the anticipations they generated are confirmed by further observation.
- Each channel periodically selects tasks with the highest priority and sends them to the overall experience buffer.

The *overall experience* buffer is similar to the I/O channels, except that its inputs come from multiple channels, so the compound terms formed in it may contain components from different modalities. Selected tasks from this buffer will be added to the memory for long-term storage and processing. Besides temporal composition, this buffer also selectively adds higher-order statements that explicitly state the relationship between the system and a statement in the buffer. For example, if statement S is selected, the conclusion may be "I am informed S," "I see S," etc. (in Narsese).

The memory of OpenNARS is a conceptual network, as described previously. In the memory, the system's experience is summarized into beliefs and desires, representing what the system considers to be the case and what it wants to be the case, respectively. The content of beliefs and desires are all Narsese statements, with an associated truth-value or desire-value, respectively, indicating the extent of the belief or desire.

In each inference cycle, a task and a belief are selected probabilistically from the existing ones in memory, and the selection is biased by the priority of the candidates. The applicable inference rules will be triggered and derive new tasks. The results of inference also include goals for certain operations to be executed, either on the (external) environment via an I/O channel, or on the (internal) memory.

The derived tasks are added into the *internal experience* buffer for selection and compound-term composing, just like the external experience, and the selected tasks will be added to the overall experience as well. Therefore, internal and external experiences of the systems are processed initially in parallel and eventually merged. The internal experience of the system also includes a record of perceived internal events, such as the arrival of high-priority input, the execution of the system's operations, and the "feelings" reported by the evaluators about the system's status, which will be explained in the following.

2.3. *Self-awareness and self-control*

As stated in [Wang and Hammer (2018)], the perception process in NARS is different from the mainstream approach in three major aspects:

- The objective of perception is to summarize the system's *subjective* experience, not to model the objective world. The rationale of this position has been justified earlier in the article.
- Perception is *active*, in the sense that the process is normally initiated by the system to accomplish its tasks, rather than always triggered by external stimuli.
- Perception is *unified* with cognition, as the processes are carried out mostly by the same rules, and controlled by the same policies.

The above features also apply to the system's knowledge obtained from its internal experience.

NARS' knowledge about its (external and internal) environment is often represented as *sensorimotor contingency*, that is, the difference in perception made by an operation, in the form of "when *pattern-a* is perceived, if the system takes *action-b*, it will *perceive pattern-c*." Under AIKR, such a statement may have both positive and negative evidence and is revisable by new evidence. The system's existing knowledge (or beliefs) may contain (explicit or implicit) contradictions, and cannot solve all problems to the system's satisfaction. This treatment is fundamentally different from that in most existing cognitive architectures, where explicit separations are made between declarative and procedural knowledge, as well as between perception and action, and the knowledge is implicitly assumed to be independent of the system and is sufficient to solve all problems given to the system [Laird *et al.* (2017)].

Even given their overall similarities, NARS' knowledge about its inside is still different from its knowledge about the external environment. This difference starts in the sensors and actuators, which provide semantic primes for knowledge representation.

As a general-purpose system, NARS has no built-in sensorimotor mechanism to interact with the outside environment. Instead, many peripheral devices (both hardware and software) can be connected to NARS, as long as their executable commands are registered in a channel that is equipped to convert the input/output data to and from Narsese. To be specific, each operation of NARS contains a channel identifier, an operator, and a list of input/output arguments that provide parameters for the operator, as well as hold return values (immediate feedback). The consequences of the operation may also be accepted later. Some of the operations change the outside environment in various ways, while some others do not, but only collect information, though all of them generate sensorimotor contingency.

On the other hand, NARS' inward sensorimotor mechanism is built-in as part of its design and is realized by a set of *mental operations* [Wang (2013)]. Sharing the same format as the outward ("physical") operations, a mental operation uses the special term *SELF* in place of channel identifier, indicating the system itself (rather than a peripheral device) is responsible for the actual realization of the operation [Wang *et al.* (2018)].

Mental operations provide self-awareness and self-control to the system. As described previously, the inference rules of NARS are predetermined, but when a task needs to be processed, there may be no predetermined algorithm that specifies the selection of premises and rules step-by-step. Instead, the premises are selected by an automatic resource allocation mechanism and some policies of priority calculations that decide at the run time according to the accumulated experience and the current situation. In this way, NARS handles novel tasks without exhaustively exploring all possibilities.

This *automatic* process is general and flexible, though not powerful enough to deal with complicated tasks. The current design also supports *deliberative* processes, in which the system explicitly reasons, plans, and decides what to do on a task, step by step. Such a *dual-process* has been discovered by psychologists in human cognition [Evans and Frankish (2009)], as the mind can neither manage all the details in the brain nor can only depend on free association when dealing with its tasks. The same strategy can be used by AI systems. Under restrictions on knowledge and resources, it is impossible for an AI system to plan all of its activities. On the other hand, to only depend on predetermined algorithms cannot adapt to an unknown or changing environment. In NARS, the built-in resource allocation mechanism provides a default control mechanism, which is supplemented by learned self-control skills.

For the system to learn when to use which mental operation, it needs to know the previous internal events, including the operations executed (routinely, accidentally, or deliberately) and their preceding and following events. In NARS, it means the relevant events must enter the system's internal experience, so as to be explicitly considered in the reasoning process by the rules responsible for the forming of sensorimotor contingencies. Whether to execute a mental operation is a decision made by the system's decision-making rule (based on the evidence provided by the other rules), rather than by the attention allocation mechanism.

Besides the operations that form the system's inference routine, there are also operations that evaluate the system's overall status. OpenNARS periodically makes four system-level evaluations [Wang *et al.* (2020)]:

- **Satisfaction**: the extent to which the current situation meets the system's desires,
- **Alertness**: the extent to which the system's knowledge is insufficient in the current situation,
- **Busyness**: the extent to which the system's time resource is insufficient,
- **Well-being**: the extent to which the system's "body" functions as expected.

Among them, the optimal values of *Satisfaction* and *Well-being* are at the high end, while those for *Alertness* and *Busyness* are at the middle. When the current value is sufficiently far from optimal, a *feeling* operation is triggered to report the issue in the system's internal experience. The system can also deliberately check the evaluations via operations and get positive or negative feelings, which is the starting point of the emotion mechanism [Wang *et al.* (2016)].

These feelings play an important role in the system's attention allocation. For example, when the system is satisfied, it usually wants to keep the current situation, rather than change it; when the system is alert, it will pay more attention to the new observations, rather than consider its long-term goals; when the system is busy, the low-priority tasks are ignored; when the system's body fails to carry out certain functions, the system will be eager to fix the problem. Like in a sensorimotor channel, NARS also acquires contingencies among its mental operations through experimenting and reasoning, to learn the preconditions and consequences for each operation it can take in its thinking process. While the automatic process is innate and remains mostly unchanged over time, the system's self-control skills grow with its experience and play a larger and larger role in problem-solving [Wang (2012a)], and the meaning of the *SELF* concept is enriched in the process, too [Wang *et al.* (2018)].

As a dual-process system, NARS is only aware of some significant internal events and controls some major operations, while leaving everything else to the innate mechanisms that are triggered by contextual factors. Consequently, the system cannot explain accurately how some beliefs are formed (they will be considered as "intuitions"), nor how a specific path of processing is chosen (they will be considered as "inspirations"). The automatic process and the controlled process are not clearly separated in function, as they use the same inference rules, and whether an event is in the system's attention can be a matter of degree. A neglected concept may get attention and explains a previous intuition or inspiration to an extent.

The set of mental operations is disjoint from all possible sets of physical operations. Even if a peripheral device can obtain all the data processed in the mental operation, it cannot process the data in the same way outside the processing context of the mental operations (provided by the contents of the memory and buffers of NARS), or cause the same effects in NARS. Unlike external perception that must go through a series of abstractions, internal perception starts at the term level, though there is still a conceptual hierarchy formed by compound terms, as the system gradually forms opinions about how its own mind works, using more and more general concepts to describe its thinking process and to understand the other minds analogically.

Consequently, the processes within NARS can also be described from two perspectives: of NARS itself (first person) and of an observer (third person). In the terminology of NARS, the former includes (executable) *operations* and the latter only contains (non-executable) *events*. The two perspectives of a process are correlated, though are expressed at different levels of abstraction using different vocabularies. They cannot be fully converted into each other, since there is no one-to-one mapping between their concepts.

3. Comparisons and Implications

In the following, the above theoretical positions and technical designs are compared with some other theories and approaches to consciousness, and their implications are discussed.

3.1. *In philosophy*

A highly influential theory on consciousness was proposed in [Chalmers (1995)] and his other writings. For the current discussion, I summarize his opinions roughly into four major conclusions:

(1) There is an explanatory gap between the *phenomenal* (or *experiential*) aspect and the *functional* (or *access*) aspect of consciousness.

(2) It is possible (at least in theory) for something (like a zombie) to have functional consciousness but no phenomenal consciousness.

(3) The explanation of functional consciousness is relatively easy, but it is hard to explain how phenomenal consciousness arises from a physical basis.

(4) The solution to the problem is the "double-aspect principle," which postulates that "there is a direct isomorphism between certain physically embodied information spaces and certain phenomenal (or experiential) information spaces."

As explained previously, I agree with the distinction between the phenomenal and functional aspects of consciousness and see it as similar to the first-person and third-person perspectives of the thinking process, respectively. Because I consider them different aspects of the same underlying process, it is impossible to have one without the other. Consequently, I do not consider zombies a valid possibility, as the lack of phenomenal consciousness means that certain information fails to be obtained and processed by the agent, which will eventually lead to different behaviors. The phenomenal aspect of consciousness is not independent of the functional aspect, as an option that the agent may have *accompanying* the functions. Instead, the phenomena are intrinsic features of consciousness, together with the functions. Though there is no exact isomorphism between the events of the phenomenal and functional aspects, there is still a rough correlation between them overall.

To me, the "hard problem" is based on a misconception. Since subjective experience does not arise from a physical basis, there are no "how" questions about it. As I explained previously, the so-called "physical basis" does not consist of objective facts, but inter-subjective descriptions, so it and subjective experience are descriptions of the same process at different levels of abstraction and from different perspectives. There is no causal relation between the two.

There are people who think to understand consciousness means to explain it in neuroscience or even physics, though Chalmers himself explicitly rejected reductionism. I agree with Chalmers that experience should be taken as fundamental. However, I do not agree with his "naturalistic dualism," according to it only a physical theory and a psychophysical theory are needed to describe the world, and other theories, like that from biology, contain no fundamental principle. As explained previously, in my opinion the differences among these theories are quantitative, rather than qualitative. Furthermore, besides the physical level and psychophysical level, there are other valid levels of abstraction on which scientific theories can be constructed.

My claim that every description is describer-dependent is not a form of mind–body dualism that assumes that mental processes and physical processes both exist, or that the mind and the body are separate entities. It is closer to the so-called "indirect realism" and "epistemological dualism" by assuming the same world can have observer-dependent descriptions, though these descriptions are correlated to various extents, and none of them is completely arbitrary or delusional, as they are all about, and therefore restricted by, the same underlying process.

There are other strong criticisms of reductionism coming from neuroscience (such as [Freeman (1999)]) and physics (such as [Anderson (1972)]) that contain arguments I agree with, though my argument is most from the considerations of cognitive science and AI, where the dependency of beliefs and knowledge on the cognitive capability and process of the system forming or holding them can no longer be omitted, like in many other fields. Furthermore, using NARS we can show how a "subjective belief" evolves into an "objective fact," step by step.

Another major philosophical issue related to this discussion is the long-lasting contrast between determinism and free will. Most people believe that "Our world is either deterministic or nondeterministic, but not both" [Müller *et al.* (2019)], though there are also suggestions like "neuronal stochasticity may be a main prerequisite to keeping the brain in a flexible state, also for decision making" so that "It seems again to be the combination of chance and necessity determining the functions of life" [Braun (2019)].

As I have rejected the postulate that there is an objective description of anything, whether a process is "deterministic" depends on the observer or describer. The same process can be deterministic to one agent, but nondeterministic to another one, depending on their knowledge, levels of description, requirements on the accuracy and reliability of the conclusion, etc.

Under AIKR, I certainly do not accept the Newton-Laplace version of determinism, but still agree that as we know more and more about how the brain/mind complex works, we can take certain aspects or events as deterministic to the extent that relevant knowledge can be used to build a computational model that makes more reliable predictions, even if the stochasticity of neural activities are taken into account. In principle, we may even be able to know how a person will act before the person is conscious of the decision herself/himself.

On the other hand, from the viewpoint of the person who is making the decisions, such predictions are impossible given AIKR. As far as the person sees the situation, she/he is the one who is about to make a decision freely and is responsible for its consequences. Therefore, free will is typically from a first-person perspective, while determinism is from a third-person. To me, there is no contradiction to accepting both determinism and free will as different perspectives or seeing the world as both deterministic and nondeterministic, with proper definitions of the concepts involved. I see the property "deterministic" as a relation "can be determined by someone" without explicitly specifying the "someone" or the reliability of the conclusion.

3.2. *In psychology*

Among the related psychological issues, the most prominent one is the distinction between conscious and unconscious processes in thinking.

On this topic, there is a well-known theory since the work of [James (1890)] that human thinking consists of both an automatic (unconscious) process and a controlled (conscious) process, where the former happens everywhere and without special effort, while the latter is focused and demands attention. The concept of the unconscious mind was popularized by [Freud (1965)] and used to explain various psychological phenomena. The related issues have been studied by many researchers, though there are still major questions to be answered [Cleeremans (2014); Evans and Frankish (2009)]. Instead of surveying the literature, here I only explain how these questions are answered in NARS.

As explained previously, NARS is designed with a similar distinction, though the details are not identical to what happened in the human mind. I see consciousness mainly as the system's awareness and control of itself, with a functional aspect and a phenomenal aspect. What is described above is the functional aspect of NARS. Though still in a preliminary stage, NARS does have knowledge about itself that is acquired from its experience, and this knowledge has impacts on the system's behavior. In this sense, the system knows what it is doing, though this knowledge is limited, and can even be wrong, judged by its future self or a well-informed observer.

As far as actions are concerned, the "conscious vs. unconscious" distinction largely overlaps with the "controlled vs. automatic" distinction among the processes in NARS. As described before, there is a default mechanism managing the routine reasoning/learning activities of NARS, so the system can carry out these activities without explicitly thinking about them, as someone doing free association or daydreaming. On the other hand, the mental operators allow the system to deliberately control its own thinking process to a certain extent. Conscious thought is closely associated with controlled thinking, which adds flexibility to the unconscious/automatic processes, though the latter is more efficient and stable, just like the relation of these processes in the human mind. Of course, I am not claiming that the consciousness/unconsciousness distinction in AI systems will be identical to that in the human mind, but that they play similar roles.

For example, if an input event E_1 triggered an output event E_2, the system is conscious about these two events only when their relation becomes part of the system's experience, such as in the form of implication judgment "E_1 implies E_2." Under AIKR, NARS cannot know every event within itself, nor can it control all of them. Consequently, self-awareness and self-control happen selectively. For example, if event E_3 merely enters the system's input channel, it is not necessarily consciously perceived; if its priority is high enough it may be picked up to generate event "E_3 is happening," and becomes conscious. For this reason, our approach is similar to the opinion that conscious processes are higher-order [Carruthers (2016)] meta-cognition [Cox (2005)].

To be more specific, it is reasonable to say that NARS is conscious about part of the content of the overall experience buffer, including the recently perceived events in the external and internal environment, as well as the recent conclusions and decisions made by itself. Since items in the buffer have different levels of priority and obtain different amounts of attention, "conscious" should be taken as a matter of degree. However, it does not mean that every process in the system is conscious. Instead, most processes are completely unconscious, meaning that they happen without corresponding representation in the system's overall experience, even though their cumulative effects may get into it when they become significant enough to be noticeable.

Since self-awareness and self-control are both provided by mental operations, so the scope of consciousness in NARS is restricted by the set of mental operators. If a version of NARS has a larger set of mental operators, its consciousness will be richer and more complicated than another version of NARS that has a smaller set of mental operators. Since we have not seen any reason for all intelligent systems to be equipped with the same mental operator set, they may have different scopes (or levels) of consciousness.

In psychology, many discussions of consciousness are actually about *human* consciousness, as they are closely associated with the underlying neural or biologic processes [Dehaene (2014)]. I do not agree with claims like "AGI is attempting to reproduce all human behaviors linked with intelligence" [Gamez (2008)], because I think it is neither necessary nor possible to reproduce *human behaviors*, which depend on human-specific features. On the contrary, the *principles* behind human behaviors, like "adaptation under AIKR" and "uniformly treating external and internal environments" can be followed in computer systems [Wang (2019a)].

3.3. *In artificial intelligence*

At present, it is still quite common to find extreme opinions with respect to the relationship between AI and consciousness: there are people who completely dismiss the possibility of AI systems being conscious, while on the contrary, there are people who consider everything conscious to different degrees. A major reason for this diversity of opinions is that the concept of consciousness has been used with very different meanings.

Nevertheless, there are also researchers taking a position somewhere between these two extremes. To them, ordinary computer systems are not conscious, though it is possible to build conscious AI and robots using some new theories and techniques. This type of opinion can be found in [Chella *et al.* (2008); Baars and Franklin (2009); Cleeremans (2014); Perlis and Brody (2019)]. Unlike philosophers and psychologists, the focus of AI researchers is the cognitive functions of consciousness, though some of them also addressed its phenomenal aspects.

A well-known theory of consciousness, the Global Workspace Theory (GWT), has been realized in an AGI system LIDA [Baars and Franklin (2009)]. According to this theory, "consciousness is associated with a global workspace — a fleeting memory capacity whose focal contents are widely distributed ('broadcast') to many unconscious specialized networks" [Baars and Franklin (2009)]. As this approach makes the conscious vs. unconscious distinction by the *location* (type of memory) where the processing happens, it is different from how this distinction is made in NARS. However, there is still a strong correlation between these two approaches.

In the architecture of NARS (as shown in Fig. 1), the overall experience buffer, the internal experience buffer, plus the storage space in the inference engine, play the role of a "global workspace," while the channels and concepts mostly correspond to the "unconscious specialized networks," because the information stored in the latter is partitioned into specific concepts or sensor/actuator, while the information in the former is shared across the whole system. The overall experience buffer holds the tasks to be processed, the internal experience buffer contains new tasks just generated, and the inference engine is where the processing happens. In this way, NARS shares the feature stressed by [Baars and Franklin (2009)]: "A striking feature of GWT is that it accounts for both the massive parallel processing of the human brain, most of which is not conscious (i.e., not reportable) and the surprisingly narrow moment-to-moment capacity of the conscious stream of thought."

The difference between the two systems is that in NARS being in the above workspace is only an approximate *necessary* condition for a data item to be conscious, but not a *sufficient* condition for it, as data in those buffers will not be accessed if their priority values are not high enough. Even the necessary condition is only an approximation, as the other types of memory may also have tasks under conscious processing. Furthermore, there is no "broadcasting" process, as message-passing is carried out by the inference tasks, which are only sent selectively to the corresponding concepts to process, not to other concepts.

There are several approaches that treat consciousness as higher-order or meta-cognition [Chella *et al.* (2008); Chatila *et al.* (2018); Reggia *et al.* (2018); Kwiatkowski and Lipson (2019); Perlis and Brody (2019)], which are similar to the position we take in NARS. Since each of these systems has different architecture and mechanism, what is considered "higher-order" is quite different, so it is not easy to compare them with each other. What I want to highlight about NARS is its *unified* representation (in Narsese) and processing (using NAL) of first-order and higher-order knowledge, as well as of first-person and third-person perspective. This unification makes the design more consistent, efficient, and elegant.

NARS also unifies conscious and unconscious processes, in the sense that the same representation language and inference rules are used for both. Their differences are more in attention allocation and whether the process is automatic. Since NARS is a reasoning system, many people may consider it a traditional "symbolic AI." However, with experience-grounded semantics, reasoning-based learning, dynamic resource allocation, etc. NARS actually shares many properties with connectionist models [Smolensky (1988)], including the features listed in [Cleeremans (2014)] — "active representation, emergent representation, graded processing, and mandatory plasticity," which are argued to be necessary for unconscious processing. Consequently, the challenge raised in [Cleeremans (2014)], "how the symbolic representations characteristic of conscious information processing can emerge out of the subsymbolic representations characteristic of unconscious information processing" does not apply to NARS, as the "symbolic vs. subsymbolic" distinction no longer exists there, so it cannot be aligned with the "conscious vs. unconscious" distinction, which does exist in NARS.

In NARS, the inward and outward sensorimotor mechanisms follow the same principles, though the sensors and actuators are different. Consequently, there is also an explanation gap. In principle, NARS can describe its internal processes from both first-person and third-person perspectives. These two descriptions are not isomorphic, though roughly correlated or associated, as discussed previously. It is quite likely that the system will also be puzzled by its own mind-body problem, or even believe that only itself is conscious because it cannot feel what it is like to be anyone else. In this case, to merely share input data among AI systems will not be enough, because the first-person perspective also depends on the mental structure that has been formed within the agent.

Finally, there is the issue of moral and ethical consequences of machine consciousness. Like all major technical breakthroughs, progress in AGI research may produce both positive and negative effects on human society, and we should do our best to avoid undesired results. However, I do not see much evidence for the popular belief that as soon as an AI becomes conscious, it will inevitably rebel against human beings and attempt to dominate the world. On the contrary, when raised properly, a conscious AGI may be safer and more reliable in a complicated situation. Since this is a big topic in itself, detailed discussion will be left for other publications. In general, I consider consciousness a necessary feature for an advanced intelligent system and believe that human society will benefit from progress in the study of machine consciousness, both by getting a better understanding of our own thinking process and by obtaining more powerful techniques to solve the problems we are facing.

4. Conclusion

By my definition [Wang (2019a)], *consciousness* is not a necessary feature of *intelligence*, but that of an advanced type of intelligent system, where the system can perceive and act on its own cognitive processes.

To me, consciousness mainly refers to the self-awareness and self-control of a cognitive system, or agent, and its phenomenal and functional aspects correspond to its first-person and third-person descriptions, respectively. Since these two types of description consist of different concepts and no perfect translation exists between them, there is an explanation gap. However, it does not mean that consciousness cannot be studied scientifically, nor that it cannot appear in AI systems.

Consciousness can be specified abstractly, and at this level, the mechanism in the human brain and advanced AGI systems can follow the same principles, though the concrete contents of consciousness, as well as the separation between conscious and unconscious processes, will be different from system to system, due to their different natures and nurtures.

The constructive model of consciousness in NARS is built by abstracting the functions observed in the human mind to a level of description where they become desired and feasible for computer systems that are designed to work in certain types of environments. In this way, I give consciousness a functional explanation without involving concepts at the neural level, not to mention the quantum level.

Based on the understanding that functional and phenomenal consciousness correspond to the same underlying process, the above cognitive functions also credit NARS with phenomenal consciousness, though it is still very simple and preliminary. I agree with the subjective and private nature of this aspect of consciousness, which means that we will never get the feeling of what it is like to be a NARS. However, it is not a reason to deny its consciousness, since NARS has the feeling of what it is like to be a NARS — actually, it is all that it can feel.

Given the complexity of all the "self" related phenomena [Hofstadter (1979)], the work presented in this article is just the first step towards a satisfactory explanation of consciousness. Even so, its constructive nature makes the ideas directly testable, so as to make it possible for them to be gradually refined and enriched into a competitive theory of intelligence, cognition, and consciousness.

Acknowledgments

I thank Fanji Gu and Karl Schlagenhauf for their profound discussions on the related topics, and the members of the OpenNARS team for their contributions to the project.

References

Anderson, P. W. (1972). More is different, *Science* **177**, 4047, pp. 393–396.

Aristotle (1989). *Prior Analytics* (Hackett Publishing Company, Indianapolis, Indiana), translated by R. Smith.

Baars, B. J. (1988). *A cognitive theory of consciousness* (Cambridge University Press, Cambridge).

Baars, B. J. and Franklin, S. (2009). Consciousness is computational: The LIDA model of global workspace theory, *International Journal of Machine Consciousness* **1**, pp. 23–32.

Blackmore, S. (2004). *Consciousness: An Introduction* (Oxford University Press).

Braun, H. A. (2019). Stochasticity versus determinacy in neurodynamics – and the questions of the 'Free Will', in *Proceedings of the 7th International Conference on Cognitive Neurodynamics (ICCN2019)*.

Carruthers, P. (2016). Higher-order theories of consciousness, in E. N. Zalta (ed.), *The Stanford Encyclopedia of Philosophy*, Fall 2016 edn. (Metaphysics Research Lab, Stanford University).

Chalmers, D. J. (1995). Facing up to the problem of consciousness, *Journal of Consciousness Studies* **2**, 3, pp. 200–219.

Chatila, R., Renaudo, E., Andries, M., García, R. O. C., Luce-Vayrac, P., Gottstein, R., Alami, R., Clodic, A., Devin, S., Girard, B., and Khamassi, M. (2018). Toward self-aware robots, *Frontiers in Robotics and AI* **5**, doi: 10.3389/frobt.2018.00088.

Chella, A., Frixione, M., and Gaglio, S. (2008). A cognitive architecture for robot self-consciousness, *Artificial Intelligence in Medicine* **44**, pp. 147–154.

Chella, A. and Manzotti, R. (2012). AGI and machine consciousness, in P. Wang and B. Goertzel (eds.), *Theoretical Foundations of Artificial General Intelligence* (Atlantis Press, Paris), pp. 263–282.

Cleeremans, A. (2014). Connecting conscious and unconscious processing, *Cognitive Science* **38**, pp. 1286–1315.

Cox, M. T. (2005). Metacognition in computation: A selected research review, *Artificial Intelligence* **169**, pp. 104–141.

Dehaene, S. (2014). *Consciousness and the Brain: Deciphering How the Brain Codes our Thoughts* (Penguin Books, New York).

Evans, J. and Frankish, K. (eds.) (2009). *In Two Minds: Dual Processes and Beyond* (Oxford University Press, Oxford).

Freeman, W. J. (1999). Consciousness, intentionality and causality, *Journal of Consciousness Studies* **6**, 11-12, pp. 143–172.

Freud, S. (1965). *The Interpretation of Dreams* (Avon Books, New York), translated by James Strachey from the 1900 edition.

Gamez, D. (2008). Progress in machine consciousness, *Consciousness and Cognition* **17**, pp. 887–910.

Hofstadter, D. R. (1979). *Gödel, Escher, Bach: an Eternal Golden Braid* (Basic Books, New York).

James, W. (1890). *The Principles of Psychology* (Henry Holt and Company).

Kowalski, R. (1979). *Logic for Problem Solving* (North Holland, New York).

Kwiatkowski, R. and Lipson, H. (2019). Task-agnostic self-modeling machines, *Science Robotics* **4**, 26, p. eaau9354.

Laird, J. E., Lebiere, C., and Rosenbloom, P. S. (2017). A standard model of the mind: Toward a common computational framework across artificial intelligence, cognitive science, neuroscience, and robotics, *AI Mag.* **38**, 4, pp. 13–26.

Müller, T., Rumberg, A., and Wagner, V. (2019). An introduction to real possibilities, indeterminism, and free will: three contingencies of the debate, *Synthese* **196**, pp. 1–10.

Nagel, T. (1974). What is it like to be a bat? *Philosophical Review* **83**, October, pp. 435–50.

Newell, A. and Simon, H. A. (1976). Computer science as empirical inquiry: symbols and search, *Communications of the ACM* **19**, 3, pp. 113–126.

Perlis, D. and Brody, J. (2019). Operationalizing consciousness, in *Papers of the AAAI Symposium "Towards Conscious AI Systems", Stanford, CA, March 25–27, 2019.*

Reggia, J. A., Katz, G. E., and Davis, G. P. (2018). Humanoid cognitive robots that learn by imitating: Implications for consciousness studies, *Frontiers in Robotics and AI* **5**, doi:10.3389/frobt.2018.00001.

Smolensky, P. (1988). On the proper treatment of connectionism, *Behavioral and Brain Sciences* **11**, pp. 1–74.

Wang, P. (2005). Experience-grounded semantics: a theory for intelligent systems, *Cognitive Systems Research* **6**, 4, pp. 282–302.

Wang, P. (2006). *Rigid Flexibility: The Logic of Intelligence* (Springer, Dordrecht).

Wang, P. (2011). The assumptions on knowledge and resources in models of rationality, *International Journal of Machine Consciousness* **3**, 1, pp. 193–218.

Wang, P. (2012a). Solving a problem with or without a program, *Journal of Artificial General Intelligence* **3**, 3, pp. 43–73.

Wang, P. (2012b). Theories of artificial intelligence – Meta-theoretical considerations, in P. Wang and B. Goertzel (eds.), *Theoretical Foundations of Artificial General Intelligence* (Atlantis Press, Paris), pp. 305–323.

Wang, P. (2013). *Non-Axiomatic Logic: A Model of Intelligent Reasoning* (World Scientific, Singapore).

Wang, P. (2019a). On defining artificial intelligence, *Journal of Artificial General Intelligence* **10**, 2, pp. 1–37, doi:10.2478/jagi-2019-0002.

Wang, P. (2019b). Toward a logic of everyday reasoning, in J. Vallverdú and V. C. Müller (eds.), *Blended Cognition: The Robotic Challenge* (Springer International Publishing, Cham), pp. 275–302, doi:10.1007/978-3-030-03104-6_11.

Wang, P. (2020). A constructive explanation of consciousness, *Journal of Artificial Intelligence and Consciousness* **7**, 2, pp. 257–275.

Wang, P. and Goertzel, B. (2007). Introduction: Aspects of artificial general intelligence, in B. Goertzel and P. Wang (eds.), *Advance of Artificial General Intelligence* (IOS Press, Amsterdam), pp. 1–16.

Wang, P. and Hammer, P. (2018). Perception from an AGI perspective, in M. Iklé, A. Franz, R. Rzepka, and B. Goertzel (eds.), *Proceedings of the Eleventh Conference on Artificial General Intelligence*, pp. 259–269.

Wang, P., Hammer, P., Isaev, P., and Li, X. (2020). The conceptual design of OpenNARS 3.1.0, Tech. Rep. 11, AGI Team, Temple University, Philadelphia, USA.

Wang, P., Li, X., and Hammer, P. (2017). Self-awareness and self-control in NARS, in T. Everitt, B. Goertzel, and A. Potapov (eds.), *Proceedings of the Tenth Conference on Artificial General Intelligence*, pp. 33–43.

Wang, P., Li, X., and Hammer, P. (2018). Self in NARS, an AGI system, *Frontiers in Robotics and AI* **5**, doi:10.3389/frobt.2018.00020.

Wang, P., Talanov, M., and Hammer, P. (2016). The emotional mechanisms in NARS, in B. Steunebrink, P. Wang, and B. Goertzel (eds.), *Proceedings of the Ninth Conference on Artificial General Intelligence*, pp. 150–159.

Chapter 4

On Artificial Intelligence, Consciousness and Robots

Pentti O. A. Haikonen

Department of Philosophy, University of Illinois at Springfield
One University Plaza, Springfield, IL, 62703, USA
pentti.haikonen@pp.inet.fi

Will Artificial Intelligence soon surpass the capacities of the human mind and will Strong Artificial General Intelligence replace the contemporary Weak AI? It might appear to be so, but there are certain fundamental issues that have to be addressed before this can happen. There can be no intelligence without understanding, and there can be no understanding without getting meanings. Contemporary computers manipulate symbols without meanings, which are not incorporated in the computations. This leads to the Symbol Grounding Problem; how could meanings be incorporated? The use of self-explanatory sensory information has been proposed as a possible solution. However, self-explanatory information can only be used in neural network machines that are different from existing digital computers and traditional multilayer neural networks. In humans self-explanatory information has the form of qualitative sensory experiences, qualia. To have reportable qualia is to be phenomenally conscious. This leads to the hypothesis about an unavoidable connection between the solution of the Symbol Grounding Problem and consciousness. If, in general, self-explanatory information equals to qualia, then machines that utilize self-explanatory information would be conscious. The author presents the associative neural architecture HCA as a solution to these problems and the robot XCR-1 as its partial experimental verification.

Keywords: artificial intelligence; machine consciousness; symbol grounding problem; qualia; cognitive robots

1. State of the Art

According to recent publicity, Artificial Intelligence would seem to be doing well. Advances have been reported in various areas, such as game playing, natural language translation, information search, robots, smart phones, smart assistants, and self-driving cars, to name a few. This publicity and recent movies about AI and conscious robots have created the popular impression that Artificial Intelligence is an autonomously acting unified entity that will soon, if not already, be able to think, understand and be intelligent in superior ways and even be conscious. Unfortunately, we are not there yet. Contrary to the popular perception the progress towards superior general artificial intelligence has been very slow. Contemporary AI is not an entity or agent, not even a unified style of programming, it is just a label given to various detached programs. Invented names do not imply the existence of any entities behind them.

Does a computer have an inner vision of the situation at hand? Does a computer have inner speech and inner imagery? These phenomena are popular hallmarks of thinking and the way of human cognition. Computers do not have these, and contemporary AI does not utilize these, either.

It is obvious that major advances in AI cannot be achieved without the consideration of the very idea of Artificial Intelligence. What is AI, what can be achieved with it, what are its limitations? Could or should AI be conscious? Is the research on the right tracks? These questions must be considered from theoretical and practical viewpoints.

In recent years remarkable philosophical groundwork on AI has been done by well-known researchers like B. Baars, P. Boltuc, N. Block, D. J. Chalmers, R. Chrisley, D. Dennett, S. Harnad, J. Reggia, J. Searle, A. Sloman, G. Tononi and others.

On the practical side, recent impactive works of I. Aleksander, S. Franklin, G. Hesslow, Y. Kinouchi, R. Manzotti, M. Shanahan, to name a few, are well-known. Biologically inspired robots have been studied by A. Chella, R. Brooks, B. Goertzel, P. Haikonen, O. Holland, K. Kawamura, R. Sanz, J. Takeno and others.

A building is only as strong as its foundations. The same goes for Artificial Intelligence. Therefore it is useful to inspect the early foundations of AI.

2. Early Foundations of AI

The theoretical foundations of AI go back to the 1950's and were explicitly presented by Newell and Simon [1976]. In the early days of computers and AI, Newell and Simon proposed that intelligence is based on symbol manipulation, and a physical symbol system (a computing machine) has all the necessary and sufficient means for general intelligence. Consequently, Newell and Simon argued that a physical symbol system is necessarily required whenever intelligence is to be produced. Humans are intelligent; therefore, according to this reasoning, also the human brain has to be a physical symbol system. On the other hand, a digital computer is known to be a physical symbol system; it manipulates binary word symbols by given rules, the program. Based on the Church-Turing thesis [Church, 1941] it is taken that a computer can compute everything that can be expressed as proper algorithms in the form of computer programs. (This statement is true because it is a tautology. Any proper algorithm is computable by definition.)

Newell and Simon concluded along the above reasoning that if both the computer and the brain are physical symbol systems, then they have to be computationally equivalent; a suitably programmed computer can execute every algorithm that is executed by a brain. These conclusions are also known as the Physical Symbol System Hypothesis (PSSH), which has been taken to confirm that human-level artificial intelligence can be produced computationally. However, Newell and Simon were not able to prove this hypothesis directly. Instead of this, they presented indirect evidence: There are no other hypotheses that could explain how intelligence could be produced in the brain or a machine. Fodor seconded this view by stating that physical symbol manipulation is "the only game in town" [Fodor, 1975].

However, the Physical Symbol System Hypothesis leaves open some fundamental questions. Obviously, the brain can operate as a physical symbol system, but could it also be something else, negating the computer–brain equivalence? Also, if that were the case, would a physical symbol system really have all the necessary and sufficient means for the production of general intelligence? Is thinking really mere execution of programs? So far, no artificial true general intelligence has been produced by computer programs. Would that be due to lack of trying or could the Physical Symbol System Hypothesis be invalid?

It is a fact that existing Artificial Intelligence programs are effective only in narrow area applications; they do not exhibit any general intelligence, perhaps any intelligence at all; they only produce an impression of intelligence, and not always even that. Therefore they are called Weak AI, as opposed to the currently hypothetical Strong AI (also Artificial General Intelligence, AGI), which would be comparable to human cognition, intelligence and the conscious mind.

3. Weak Artificial Intelligence

Computer programs that can generate apparently intelligent results in well-defined narrow areas are called Weak Artificial Intelligence (Weak AI). With the exception of few experiments, existing AI applications represent Weak AI.

Remarkable results have been recently reported with Weak AI. Considering this, it can be asked what would be the point of Strong AI, if, already in the near future, stacked Weak AI programs were able to produce all that is practically required. Unfortunately this will not happen. Weak AI will remain weak for a simple reason; it does not understand anything. This is a direct consequence of its fundamental way of operation.

Weak AI is based on algorithmic symbol processing. An algorithm is a sequence of rules to be followed exactly for the production of the desired outcome. Computer programs are algorithms. Algorithms have a benefit: The executor of an algorithm does not have to understand, what the rules are ultimately about. Consequently, algorithms can be executed by persons, agents and machines that do not understand the meanings of the used symbols or the actions taken. For example, a calculator operates with symbols (numbers), but does not know what is being computed. The external meanings of the numbers are not entered into the calculator.

This benefit of algorithm is also its greatest weakness: An algorithm is able to do only what it is designed to do, therefore its application area is inherently narrow. A chess-playing algorithm is not able to do anything else, no matter how trivial, it is not even aware that it is playing chess. Weak AI is not readily able to do any outside-the-box reasoning. A computer does only what the program commands it to do, not necessarily what was desired (a fact that is soon learned by a student and relearned by complex systems programmers). This, of course, could be remedied by providing a rule or a program for each and every possible situation.

If one rule were not enough, then how many rules would be needed? It turns out that a very large number of rules, perhaps even conflicting ones, would be required for the production of Artificial General Intelligence by computational brute force.

Nevertheless, do meanings really matter? Consider this: A triangle-shaped cookie has three corners How many corners remain if one corner is cut away? The answer is not two. Simple computation fails here, and the correct answer calls for the evocation of the idea of a triangle and the action of cutting one corner; meanings matter.

Real human intelligence is not rule following. On the contrary, it is something that is used when rules don't work. It is about finding and learning new responses that fit the situation. But this would call for the understanding the situation; meanings would have to be manipulated instead of the blind manipulation of symbols. This requirement leads to the so-called Symbol Grounding Problem: How to attach meanings to the used symbols.

4. The Symbol Grounding Problem

There is no understanding without meanings. In order to understand something a computer would have to operate with meanings, but this does not happen because physical symbol systems manipulate symbols, not their meanings, which are not present within the system. The external meanings are given to the products of the computations by human users, not by the computer.

However, how come then that certain already existing AI-based assistants, like Alexa and others, are able to converse in a natural language? After all, they seem to understand what has been asked, and are able to answer more or less properly. The inconvenient truth is these systems do not really understand anything. The words refer to words, not to their meanings, and tricks are used to create the illusion of meaningful conversation.

In a digital computer, symbols can be made to refer to other symbols, but not to their meanings because the meanings are not readily attached to the symbols. The meanings of the used symbols cannot be defined ultimately by other symbols within the system; they would have to be imported from the outside. But the imported meanings cannot be imported as additional symbols, as also these would call for interpretation. In pure symbol processing systems this would appear to be an unsolvable problem as illuminated by the Chinese Room thought experiment of Searle [1984, 1997], where a non-Chinese person inside a closed room answered questions written in Chinese characters solely by given rules and lists. Did the person understand what the questions were about? From the outside it appeared so, but no. The person simply did not have the necessary information, and had no way of acquiring it because also this information would have been in the form of Chinese characters, which the person did not understand. The situation does not change if the person is replaced with a computer; the meanings are not available.

This Symbol Grounding Problem manifests itself also as the "Dictionary explanation problem". In dictionaries the meanings of words are explained by other words, and these again by other words. Eventually this leads to circles, as the words to be explained will be used to explain the words that were supposed to explain the words to be explained in the first place. At the end nothing has been explained.

The Symbol Grounding Problem is also relevant to the process of thinking. In humans, thinking involves especially the flow of inner imagery and inner speech. Inner speech utilizes a vocabulary of words and also syntax, the apparent formal rules for the structure of sentences. With the help of syntax, words are able to refer to other words, but this is not enough because the words have to refer to their meanings as well.

In human thinking the meaning comes before words; authors know this, and many of us may have sometimes experienced an awkward situation, where we have the meaning, but do not find the proper words. In the human mind the Symbol Grounding Problem has been solved; we know what our thoughts are about.

Technically, the Symbol Grounding Problem involves three issues; firstly, how to generate a symbol, secondly, how to attach a meaning to it, and thirdly, how to manipulate the meanings, not only the symbols.

5. Strong AI and Consciousness

Strong AI, also known as Artificial General Intelligence (AGI), is supposed to produce human-like cognition, free thinking, reasoning and intelligent common sense without area limitations. Humans understand what they are thinking about because thinking operates with meanings. Artificial General Intelligence is also supposed to understand what it is doing, but without the manipulation of meanings that will not happen. Meanings must be incorporated, and in working AGI systems the Symbol Grounding Problem has to be solved.

The fact that meanings cannot be imported into physical symbol systems in the form of additional symbols leads to the obvious conclusion that the imported meanings must be in the form of non-symbols. Harnad [1990] and others have proposed that the Symbol Grounding Problem could be solved by grounding the meaning of symbols in non-symbolic representations of the external world, produced by sensory processes. This principle would appear to be an obvious and straightforward one, but problems remain, see for instance Taddeo and Floridi [2005]. How could a digital computer accommodate both non-symbols and symbols? It may not.

It has been argued that the Symbol Grounding Problem cannot be solved in digital computers because they can only accept information in symbolic (usually numerical) form [Haikonen, 2019]. For computers the sensory information captured by sensors like cameras, microphones etc. must be digitized into streams of binary numbers. The acquired information is now in symbolic form, but the phenomenal meanings are lost; numbers are not experiences of visions, sounds, pleasure or pain, they are nothing comparable to the human sensory experiences that are so evident to us. Naked numbers do not convey external meanings.

Recently it has been proposed that in *other than digital computers* the Symbol Grounding Problem can be solved by importing external meanings in the form of self-explanatory sensory information [Haikonen, 2019]. In humans self-explanatory forms of sensory information appear as qualia.

Qualia (singular: quale) are internal qualitative appearances of direct sensory percepts and also the appearance of the virtual percepts of all mental content like inner speech, imaginations, pain and pleasure. For instance, the quale of lone laser light with 650 nm wavelength is red, the quale of a circle is a round pattern, the quale of a fast air pressure vibration is a sound. The mind takes these as actual properties, not representations, of the outside world, even though actually they are sensors' neural responses to these. Qualia are self-explanatory; they appear directly as the qualities and properties of the sensed entities; no interpretation is necessary.

Cognition is more than the mere perception of qualities; it is also operation with symbols. Yet, qualia are all that is available for the mind, and therefore there must be a way in which they can be used as symbols. This can be done via association. For instance, a sound pattern may be associated with a sensory percept of an entity or action. Thereafter the sound pattern would also stand for and be a symbol for the associated entity, while its appearance would remain as the quale of the sound pattern. This process also detaches the associated percept from its temporality; the associated meaning can now be evoked by its symbol also when it is not sensorily present. This solves the Symbol Grounding Problem in the human mind; the basic meanings are imported in the form of qualia, and further meanings can be grounded in these via associative linking, leading to networks of meanings [Haikonen, 2003, 2007, 2019].

What has consciousness to do with this? Introspection shows that consciousness is based on perception; at each moment the contents of consciousness consist of reportable qualia-form percepts about the external world, the person's own body and its functions, own thoughts, inner imagery and feelings. Everything in the conscious mind has the form of qualia, and a conscious person can remember and report these percepts to itself and possibly to others at least for a while. Whenever the perception process ceases, like in deep dreamless sleep, also consciousness vanishes because there will be nothing to be conscious of. And vice versa, the flow of qualia-form reportable percepts generates the experience of consciousness. As such, consciousness is not a material or immaterial entity or agent, it is just self-reportable perception with self-explanatory qualia.

The Symbol Grounding Problem can be solved also in cognitive machines by the use of self-explanatory percepts. These percepts would have to have qualities that relate to the sensed stimuli and distinguish them from each other; therefore they would be some kind of qualia. In humans, the flow of qualia-form percepts generates the experience of consciousness. If this were the case in general, then the cognitive machines that utilize self-explanatory information would be conscious in the aforestated sense, even though their qualia would not necessarily be similar to human qualia.

This leads to the hypothesis about an unavoidable connection between true Artificial General Intelligence and consciousness: True Artificial General Intelligence has to operate with meanings, and that requires the solving of The Symbol Grounding Problem. This problem can be solved by importing the meanings in the form of non-symbolic self-explanatory information, which has the form of qualia. The flow of self-reportable qualia is consciousness.

Would it be possible to create conscious AI by computer programs? Could a computer have qualia? Qualia are subjective sensory experiences, not numbers, and therefore they are neither computable nor importable into computers. Therefore it seems that true human-like AGI cannot be achieved solely by symbolic computations and digital computers. Experiencing systems and machinery would be needed. Might biologically inspired approaches, like artificial neural networks, be the solution?

6. From Networks of Neurons to Networks of Meanings

The brain is a neural network that consists of a very large number of brain cells, neurons, and their connecting points, synapses. The neurons communicate with each other via the synapses and form large, interconnected networks. These can be modeled and realized artificially, either as computer simulations or as actual electronic hardware. The earliest neuron models were devised by McCulloch and Pitts Jr. [1943] and Rosenblatt [1958]. The principles of Rosenblatt's Perceptron model are still used in many applications.

A simple artificial neuron is a threshold device, which receives a number of input signals via artificial synapses, and produces an output signal if the sum of the weighted input signal intensities is higher than a set output threshold (hard or soft). The synaptic weight of each artificial synapse is adjustable, and can be controlled by different means.

Properly trained and adjusted synaptic weights allow artificial neurons and neuron networks to respond only to certain input signal patterns, and in this way they can be used as pattern classifiers and recognizers. The adjustment of the individual synaptic weights usually involves the iterative tweaking of the weights against each other, by repeated supervision or self-learning. For multilayer neural networks various synaptic weight adjustment algorithms exist, such as Back Propagation.

Artificial neural networks can be made to detect low level feature patterns and high-level combination patterns, and this information may be used by computer programs for the production of useful results. Is the Symbol Grounding Problem thus solved and does the hybrid combination of neural networks and a digital computer operate now with meanings? For instance, artificial neural networks may be used for the detection of words in the stream of heard speech, but the crucial question is does this lead to the capture of symbols or the capture of meanings? Will the computer program now manipulate meanings instead of symbols? Hardly. The person or sorter inside the Chinese Room has to be able to recognize Chinese symbols, that is, to separate them from each other, but this operation does not give access to the meanings that the symbols refer to.

Natural language uses words and their syntactic combinations, sentences, to name things and to describe actions and situations. In order to understand what a sentence means, the words and their syntactic combination must evoke mental qualia-based "imagery" or "experience" of the described situation, also known as multimodal mental models [Zwaan and Radvansky, 1998; Haikonen, 2003]. This "imagery" would be a kind of a sketch of what would be perceived and experienced if the situation were real. The presence of these mental models and "imagery" is not a mere hypothesis; it should be a familiar experience to anyone who has read a novel.

Thus, it is not enough simply to detect the words and the syntactic structure of the sentence; the meanings must be resolved and activated. This can be done by the associative evocation of already learned relevant information. Suitable mental "imagery" or "experiences" cannot be evoked, if they have not been acquired earlier. It should be noted that here "experiences" may also involve dynamic conditions and reactions of the experiencer.

Pattern detection and classification alone do not constitute understanding. In addition to pattern detection, another function is required, namely the associative linking of real and virtual sensory percepts, qualia. This calls for neurons that can both detect patterns and link them associatively. Examples of these neurons exist, e.g. [Haikonen, 2019].

Cognition does not arise from the linking of symbols, it arises from the linking of meanings. Artificial cognitive neural networks should not only be associative networks of neurons; they should amount to associative networks of meanings.

7. Associative Processing and the Problem of Choice

Introspection shows that thoughts do not follow each other randomly. Instead of that, one thought seems to be associated with the next one by some connection. This was noted already some 2500 years ago by the Greek philosopher Aristotle in his book *De Memoria Et Reminiscentia*. Later on various philosophers developed the theory known as associationism, which proposed that all mental processes, like thinking, learning, and memory, can be explained by the associative linking of ideas. Later on, however, it has been pointed out that even though associationism might appear to offer an obvious explanation, it also seems to have a fatal flaw, known as the Combinatorial Explosion.

The author has presented a modern variation of associationism in the form of an associative neural cognitive architecture HCA [2003, 2007, 2014, 2019]. The author rejects the proposition that full cognition could be achieved by mere associative linking. Associative linking is necessary, but it has to be augmented with match, mismatch and novelty detection, memories, feedback loops and attention control. Also an additional process is required for the circumvention of the Combinatorial Explosion problem.

Associative neural networks learn by accumulating associative connections between entities. In the course of time a very large number of connections may be accumulated. Consequently, a cue like a sensory percept, could be able to evoke a large number of different associations; for instance, a word may have many meanings. This leads to the Combinatorial Explosion problem and to the Problem of Choice; which associations and which meanings would be relevant within the framework of the current situation? Context-based frames [e.g. Minsky, 1975] might limit the possible associations and in this way they might solve the Problem of Choice. But frames have problems.

Dennett [1987] has provided an example of a frame problem: A robot enters a room to retrieve a given object. In the room there are a large number of various objects, including a bomb with a burning fuse. Which object should be attended to?

Generally, context would frame and limit the scope of choice and would solve the Problem of Choice, but in the Dennett's example something out of the context would be more important. Obviously, context is not enough. A general mechanism that is able to evaluate instantly the significance of each perceived object and situation is needed. This evaluation should be able to override the context-related attention, and focus attention on the more significant percepts. In associative neural networks this action would call for another network for attention control in the form of neural threshold control lines.

Percepts have qualia-based self-explanatory meanings and also learned associatively connected meanings. However, for the solution of the Problem of Choice another additional meaning is required, namely the emotional significance. All real and virtual percepts should have an attached good–neutral–bad significance, an emotional value that guides attention and initiates proper reactions. A robot with good–bad significance evaluation would approach good objects and would try to execute good tasks, while bad objects and actions would be avoided. In Dennett's example this significance would alert about the bomb. The good–bad significance could also be used to motivate the robot to execute desired actions also in unpredictable conditions.

How would the meanings of good and bad be established in the first place? In humans this takes place initially via the experience of pain and pleasure as well as the perception of pleasant and unpleasant tastes and odors. Also match and mismatch conditions seem to generate pleasure and displeasure; pleasure follows, when an expectation is met, and when it is not met, disappointment and displeasure follows. Novelty causes the surprise reaction. Similar mechanisms could be used in robots, both with and without supervision. These mechanisms would amount to some kind of machine emotions, not too much different from the human ones [Haikonen, 2003].

8. Cognitive Architectures for Robot Brains

The human brain, the senses and the various response systems including muscles form an integrated system that is able to think, control the body and produce actions and behavior in planned interactive and communicative ways. Autonomous robots should have a similar cognitive system. A number of architectures for artificial neural and symbolic cognitive systems have been devised, see e.g. the list of cognitive architectures by Samscnovich [2010]. Well-known examples of cognitive architectures include the Global Workspace Architecture of Baars [1997], the LIDA architecture of Franklin [Baars and Franklin, 2009] and the Communications Infrastructure architecture of Shanahan [2010].

A cognitive architecture is an arrangement that combines various subsystems, such as sensory perception modalities, memories and physical response modules (effectors) into a system that is able to produce human-like cognition and possibly even consciousness. Three main types of proposed cognitive architectures can be distinguished: Symbolic processing models (program), sub-symbolic processing models (neural network based) and the combination of these (hybrids). Nowadays also neural networks are usually realized as computer programs, and therefore it can be asked: Is there any real difference between these approaches? The answer is: There is no fundamental difference when done in this way; neural networks are here only another style of numerical computation. Therefore, another, completely different system approach is necessary.

Cognitive systems that understand must operate with meanings. This calls for a perceptual system with multimodal sensors that produce self-explanatory percepts and sensations. A proper cognitive architecture would be a neural one, which utilizes non-numeric self-explanatory percepts directly as such and also in symbolic ways with additional associated meanings, like the author's neural associative architecture HCA.

9. The HCA Architecture

The HCA (Haikonen Cognitive Architecture) is an all-neural, non-numeric, non-digital associative system model that integrates multimodal perception, cognitive processing and response generation for the control of embodied robots [Haikonen, 1999, 2003, 2007, 2019]. HCA is biologically inspired, and it processes information in the style of the human mind with inner speech, inner imagery, emotions and grounded meanings.

As a perceptual system, HCA conforms to what has earlier been said here about self-explanatory sensory information and qualia. Instead of numbers, HCA operates with non-numeric spatial and temporal signal patterns. Initially these signal patterns are sensors' preprocessed signal responses to the corresponding stimuli. This preprocessing extracts elementary features such as shapes, sizes, color, change, sound frequencies, rhythms, etc. This is rather similar to the distributed representations of Hinton [Hinton *et al.*, 1986], where an entity is represented by a combination of elementary feature signals. This method facilitates the recognition of an entity also when its percept only resembles the actual entity. Recognition of entities by few elementary feature signals takes place also in human perception; consider for instance caricatures and cartoons.

Elementary features appear in the HCA as non-numeric spatial and temporal signal patterns, produced as the sensors' responses to stimuli; a certain set of signal patterns is generated by a certain set of stimuli. Variable signal intensities can be used, but these do not change the meaning of the signal pattern; for instance the loudness of a spoken word does not change the basic meaning of the word, while its importance may be changed.

From the system's point of view a spatio-temporal neural sensory signal pattern is self-explanatory; for the system the signal pattern is the detected entity, not its symbol or a representation calling for interpretation. The HCA can memorize and associate these patterns with each other and use them as mental content that can be evoked also internally, without the sensory presence of the actual entities. This allows imagination, the virtual manipulation of external entities.

The basic operation mode of the HCA is the manipulation of non-symbolic neural activity patterns. However, associative processing allows also the association of various signal patterns with each other. In this way signal patterns may become symbols for other patterns. This facilitates the seamless transition to symbolic processing and natural language, which would allow the verbal description of situations and conditions.

The HCA is able to generate various cognitive responses, such as thoughts, imaginations and commands for physical actions. These are produced by "sub-conscious" associative processes that at each moment may produce a number of results. The most fitting results are selected by winner-takes-all thresholds.

The HCA, like the brain, is not able to perceive the products of the internal associative processes directly, as sensory perception is the only perception process. Therefore the system's responses, such as imaginations, inner speech and envisioned actions, must be returned into virtual sensory percepts. This is done by internal feedback loops with additional benefits. Consequently, the basic subsystem of the HCA is the so-called perception/response feedback loop. Somewhat similar perception feedback loops have been proposed also by Chella [2007, 2008], Hesslow [2002] and Steels [2003].

The perception/response feedback loops of the HCA are neural network assemblies that utilize special associative neurons (Haikonen neuron) and neuron groups, see [Haikonen, 1999, 2007, 2019]. These neurons and neuron groups differ from the traditional artificial ones; their synaptic weights are not tweaked against each other by any synaptic weight adjusting algorithms. Instead, modified Hebbian correlative learning is used, and each synapse learns and operates independently. See Fig. 1.

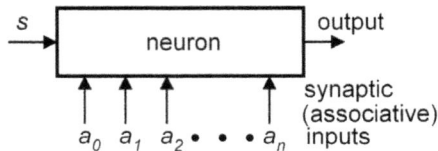

Fig. 1. The Haikonen two-function neuron.

The Haikonen neuron has one s-signal input (main signal) and a number of synaptic inputs (a_0, a_1, a_2, ...a_n). The Haikonen neuron has two functions: Firstly, it is able to learn and recognize an input signal pattern (a_0, a_1, a_2, ...a_n), and secondly, it is able to associate (link) this pattern with the s signal, and thereafter evoke this with the pattern.

In the first mode the s-signal is permanently set to 1, and the neuron learns on its own a repeating pattern of a_i signals and produces output whenever this pattern or a closely similar one is presented to it. In this mode the neuron can be used as a learning pattern recognizer.

In the second mode s is a feature signal, and $\{a_i\}$ is a signal pattern. When a certain $\{a_i\}$ pattern repeatedly coincides with the s signal, they will be associated with each other. Thereafter this $\{a_i\}$ signal pattern or a closely similar one will evoke the output signal, which will have the same meaning as the s-signal. In this mode the neuron can be used to link and associate one signal pattern with one signal. Neuron groups that associate signal patterns with signal patterns are used in the HCA architecture. (Associative memories, associators, were investigated already in 1970's, but were found to be unsatisfactory due to their limited capacity and various interference phenomena that led to mixed output patterns. In [Haikonen, 2007, 2019] it is explained how these problems can be avoided.)

Within HCA associative links can be formed between entities that are simultaneously present and entities that are made to be simultaneously present by the use of memories. Thus, spatial (parallel) and temporal (sequences) neural activity patterns can be associated with each other. These neural activity patterns may signify objects, actions, situations and also words, etc. Plain neural activity patterns can acquire additional meanings via association.

As an introduction to the workings of the HCA architecture, two cross-connected perception/response feedback loop assemblies are depicted in Fig. 2.

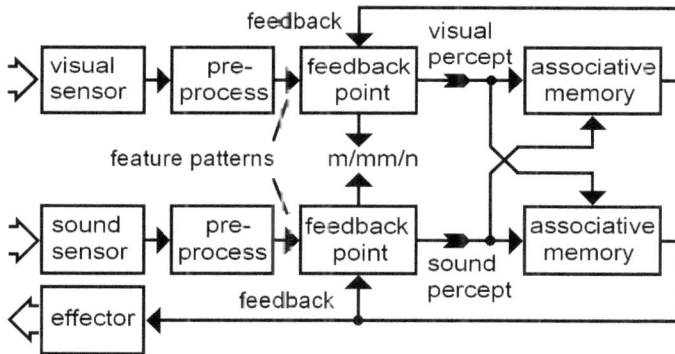

Fig. 2. Cross-connected visual and auditory perception/response feedback loops (simplified).

The visual and auditory perception/response feedback loops in Fig. 2 consist of sensors, feature extracting preprocessors, a feedback point and an associative memory. The feedback points receive both sensory signal patterns and output patterns from the associative memories and their outputs are "official percepts" of the perception/response feedback loops.

The associative memory learns spatial and temporal associations within the sensory modality and also associations with the percepts of the other sensory modality. Output patterns are evoked associatively by the inputs. These output patterns may represent expectations for sensed patterns or other mental content.

The output patterns of the associative memory are fed back to the feedback point. When the external world is being observed, attention focuses on sensory perception and the feedback is an expectation for the sensory pattern. This expectation may arise from the system's accumulated world model or it may represent, for instance, an object to be searched for. Match, mismatch and novelty conditions between actual sensory percepts and the expectations are detected and used to control attention via threshold adjustments. Also, imperfect sensory patterns may be augmented, for example, when an object is only partially seen, or a word is only vaguely heard.

When attention focuses on mental content, sensory patterns are subdued and the feedback loop will return imaginations, inner speech and other content to the feedback points. The returned signal patterns are then perceived as virtual percepts; inner speech is perceived as virtually heard speech; imaginations are perceived as virtually seen images. Also intended actions can be mentally perceived and evaluated in this way before their actual physical execution. Virtual percepts are not as vivid as real sensory percepts.

The output of the feedback point is the instantaneous "official percept", which is a part of the instantaneous contents of consciousness in the aforegiven sense and also a part of the train of thoughts. The "official percept" is forwarded to the associative memory and is also reported to the other perception/response feedback loops. The associative memory receives these reports, which may then associatively evoke various response percepts in the terms of the receiving modality. These again are returned to the feedback points of the corresponding modalities. In this way associations are looped and chained, and the train of thoughts is sustained.

Two major threshold systems operate here; the associative input thresholds, which select the accepted inputs, and the winner-takes-all output thresholds, which select the output to be fed to the feedback point and also to the effectors.

Cross-associative connections between perception/response feedback loops allow multisensory information integration and response generation. These connections also allow the seamless transition from non-symbolic to symbolic operation, which is necessary for higher cognition and natural language. For instance, the system of Fig. 2 may associate non-symbolic heard sound patterns and seen visual patterns with each other. This act gives associated meanings to these percepts; the heard sound pattern will act as a name, a symbol, for the visual pattern, and will evoke it internally as a virtual percept. And vice versa, the seen visual pattern will evoke internally the associated sound pattern, the name, and possibly a pronounced response.

The complete HCA architecture consists of a number of sensory modalities, each having a number of perception/response feedback loops. The associative memories may be quite simple, but also quite complicated, in [Haikonen, 2007] some complicated structures for language processing are given.

The Problem of Choice, attention control and motivation must be addressed in associative systems. HCA utilizes The System Reactions Theory of Emotions (SRTE) [Haikonen, 2003, 2007, 2019] as the solution. In HCA percepts are emotionally evaluated and a good, bad or neutral emotional significance is associated with them. Significance is then used to guide attention and minimize the Problem of Choice.

HCA utilizes pleasure and pain as universal motivators. In the HCA pleasure and pain are reportable dynamic system conditions that affect attention and have their typical consequences and manifestations. Pleasure and pain can be associated with entities and occurrences and can give them good-bad emotional significance.

In the HCA the emotional significance can be used as motivation: Actions that are rewarded by pleasure will be pursued, and actions that are punished by pain will be avoided, if the expected pain is worse than the possible pleasure produced by the action.

As a feasibility test, the HCA architecture has been implemented in a minimalist way in the author's XCR-1 robot.

10. The HCA-based XCR-1 Robot

The author's HCA architecture is designed as a theoretical model for robot brains that operate with meanings, "inner speech" and "inner imagery". Theoretical models are not worth much if they are not concretized, tested and verified in practice. Therefore the author has designed and built a simple non-digital neural hardware robot for the testing of the main principles of HCA [Haikonen, 2011, 2019].

The author's experimental cognitive robot, XCR-1[a] and one test target are depicted in Fig. 3. This robot is a small differential drive three-wheel robot with gripper arms and modular open chassis construction for easy testing and modification of the circuitry. XCR-1 has simple visual, auditory, touch, shock and petting sensors, and it operates with hardware-based associative neural networks in the HCA configuration; no processors, programs or analog/digital converters are used. XCR-1 is fully autonomous, not remote controlled, not connected to a computer. The block diagram of the XCR-1 robot is given in Fig. 4.

The circuitry of the XCR-1 robot consists of five associatively cross-connected modules, which are the auditory module, visual module, emotional module, gripper module and the wheel drive module.

Two separate sensory sound channels allow the detection of directions toward sound sources based on the sound intensity differences that arise from the masking of the chassis.

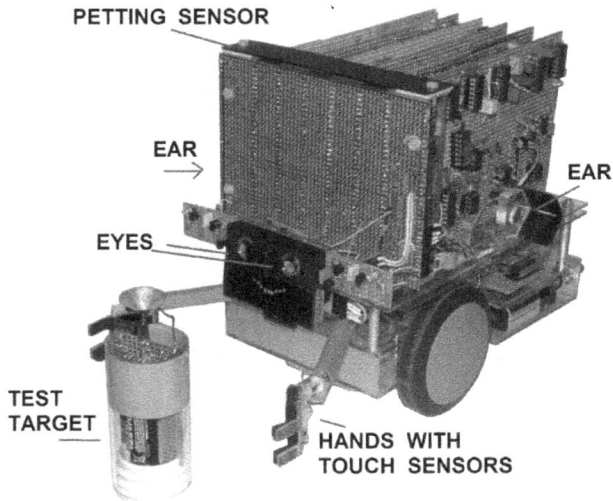

Fig. 3. The XCR-1 robot and one test target (2021 configuration).

[a]Updated demo videos of the XCR-1 robot can be seen here: https://www.youtube.com/user/PenHaiko.

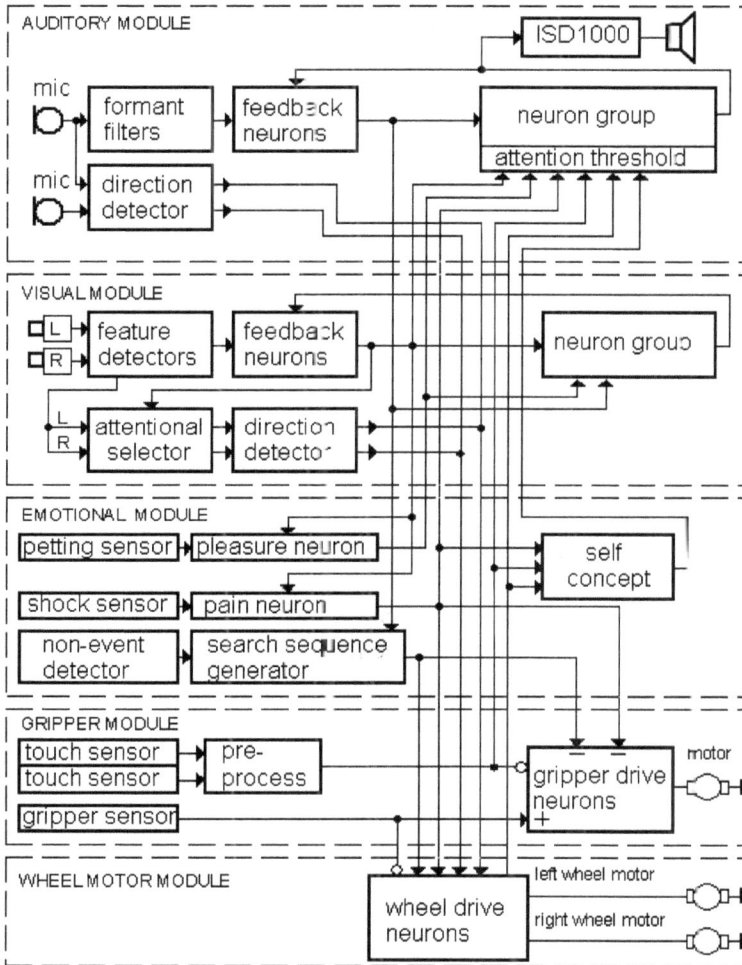

Fig. 4. XCR-1 robot block diagram (2021 version).

XCR-1 has two "eyes". These are nothing more than two large area infrared photo diodes, but they still can detect two infrared-transmitting test targets ("green" and "blue") and recognize them by the pulse rate of their infrared waves. The "eyes" can also quite accurately determine the direction toward a test target by using parallax-caused intensity differences in the received signals.

The emotional module detects petting and shock inputs, and derives its good–bad concept from these. Good and bad values can be associated with the test targets as their emotional significance.

The shock detector is bolted to the chassis, and its output is filtered for the optimal detection of mechanical vibrations of the chassis. Detected shock triggers the actual pain condition, which is a global wave of disturbance in the threshold control network. This affects attention and action, and is verbally reported.

The wheel motor module controls the two-wheel drive motors. The module receives target direction information from the auditory and visual modules. With this information the robot is able to turn and move towards the detected direction.

The XCR-1 will home in on the visually detected test target and will then capture the target between its hands. After a short while the robot will release the captured target and will back off.

The wheel motor module receives commands also from the emotional module. If nothing happens for a while, the emotional module initiates an autonomous search sequence of moving forward and turning. The search sequence stops if the robot is hit, or it hears a sudden loud noise. The search sequence stops if a target is detected, thereafter the robot will move towards that target.

XCR-1 can produce emotion-related verbal reports and cries, and it can also display its emotional states with a "smile" and "grin".

XCR-1 has inner speech with limited vocabulary. The inner speech is spoken aloud, and it consists of a running commentary about the robot's situation. Thus the robot is able to report the act of searching, seeing the "green" and "blue" targets, the touching of these and their emotional good–bad value. The vocabulary is stored in an analog audio memory chip (ISD1000).

XCR-1 feels pain in the form of dynamic disruptive system condition. Pain is verbally reportable; if the robot is hit, it will report "me hurt". If the robot sees the "green" (or "blue") target at the same time, it will immediately associate the target with "bad" and it will also report "green bad". The robot will at once back off from the target, and will later on avoid it whenever it is nearby.

XCR-1 recognizes some spoken words by so-called formant filters (narrow band audio filters), and gets also the meanings of these via cross-modality associative evocation. Spoken words "green" and "blue" evoke the corresponding virtual visual percepts in the visual module, and the word "bad" evokes the emotional value "bad" in the emotional module.

The perception/response feedback loops of XCR-1 execute also the match/mismatch detection function. Thus, if the "green" target is shown and the word "blue" is spoken, mismatch condition occurs between the sensed percept of "green" and the evoked virtual percept of "blue", and the robot says "no". If instead of "blue" the word "green" is spoken, the sensed percept and the evoked virtual percept match each other, and the robot says "yes". It should be noted that elementary feature patterns are compared here; any "pixel-wise" comparison would not work.

XCR-1 has a self-concept that is grounded on percepts about robot's own condition (pain, pleasure) and actions. XCR-1 reports the activated self-concept by the word "me". Thus the robot may report "me search", "me touch", "me hurt" or simply "me". XCR-1 is able to recognize itself in its mirror image according to the principles presented in [Haikonen, 2007b]. The seen mirror image is reported by the word "me" and a smile. This has some philosophical implications, but it should not yet be taken as a proof of self-consciousness.

Verbal teaching involves the spoken description of the conditions to be learned without the actual presence of these conditions. Verbal teaching is a good test for the HCA architecture, as it requires speech understanding, inner imagery, detection of match/mismatch conditions, the learning of associations and finally, meaningful verbal responses and later on the utilization of the learned detail in responses. All these necessitate the activation of multiple inner "images" and associative connections between various perception/response feedback loops in the HCA architecture.

Simple verbal teaching has been demonstrated with XCR-1. In this demonstration the robot is told verbally that the green test object is bad by saying "green (is) bad" repeatedly (you can say "green bad" or "green is bad". It does not matter, for the robot it is the same thing). The learning is based on evoked inner "imagery" only, the green test target is not present and pain is not induced by hitting. The trick is now to associate the inner "image" of the green object and the quality "bad" with each other. This association can only take place, if the percepts of these are available at the same time. However, the word "bad" is only spoken after the word "green" and the percepts of these words are not generated at the same time. For the required simultaneity these percepts must be sustained for a while at their corresponding modalities. This is a short-term memory action, which is an inherent property of the HCA perception/response feedback loops, and therefore no additional circuitry is required for this purpose in the XCR-1.

During the verbal teaching the sentence "green bad" leads initially to mismatch because "green" has not been associated with "bad", and thus the previous knowledge has been contradicted. There is no match, and the robot will report "no". Next teaching effort will lead to the same result and "no" report, but with repetition, the synaptic weights of the corresponding associative neurons will eventually exceed the threshold, and "green" will be permanently associated with "bad", which will now constitute the emotional significance of the green target. Next time the word "green" will evoke "bad" in the emotional module, and the word "bad" will evoke the same. Match condition occurs and the robot will now report "yes". The robot has now learned that "green" is bad. If the "green" test object is now presented, the robot will report "green bad" and it will back off and try to avoid that object. The intended association has not only been formed and learned, its meaning has also been "understood".

No doubt, the various actions and routines of the XCR-1 are trivial, but that is not the point. The important issues relate to the principles of self-explanatory perception and the cross-associative connectivity that allow the utilization and linking of meanings and the seamless transition from non-symbolic to symbolic processing.

Various experiments with the XCR-1 robot show that the basic principles of the HCA architecture are sound. Meanings can be imported into the system in self-explanatory forms and information can be processed associatively without preprogrammed rules, seamlessly in non-symbolic and symbolic ways, including inner speech. XCR-1 addresses also the issue of phenomenality; it may be the first and so far the only robot where pain is a dynamic disruptive system condition, not a symbol, even though it can be verbally reported. The question "does this dynamic disruptive system condition constitute an inner experience?", remains at the moment philosophical.

Associative parallel neural processing in the HCA style has also practical benefits: The meaning of each signal and signal pattern can be traced and the connective content-level function of each neuron and neuron group can be known; this is really helpful for the system designer. HCA is also efficient in terms of the required hardware. HCA and XCR-1 do not use software and therefore memory hardware is not needed for the storage of program lines. XCR-1 has maybe some tens of thousands of transistors (excluding the analog audio memory). A real-time simulation of XCR-1 (and especially HCA) would call for a rather complicated program with many, many program lines and a fast computer with billions of transistors to run it. And yet, the simulation would not produce the real thing.

11. Conclusions

Artificial Intelligence has not yet reached its full potential. Existing Weak Artificial Intelligence has a shortcoming that limits it power; it operates algorithmically without the utilization of meanings. Without meanings AI cannot understand anything, and without understanding there cannot be any real intelligence. This is a real challenge especially for autonomous robots, which should be able to operate in rather unpredictable everyday environments, as easily as humans do.

General Artificial Intelligence would be the ultimate state of AI. It would have to operate with meanings, but there is a problem. Existing computers operate only with symbols; therefore computer's symbols can internally refer only to symbols, not to their external meanings. Moreover, the meaning of symbols cannot be ultimately defined by additional symbols. Therefore meanings must be imported in a self-explanatory way. For this purpose novel technical approaches and system architectures are needed.

Qualia are self-explanatory information. Does this work also the other way around, would self-explanatory information necessarily have the form of qualia? If so, would neural machines experience their self-explanatory neural patterns as some kind of qualia? This leads to an interesting philosophical question. To have the flow of self-reportable qualia is to be phenomenally conscious. Therefore, would neural machines that use self-explanatory forms of information be phenomenally conscious?

The history of technology shows that more of the same is usually good. But it also shows that real breakthroughs can only come from major paradigm changes; consider for example the transition from electron tubes to transistors. A wrong way may look promising, but it will not lead to the desired destination, no matter how far you go. New revolutionary ideas and approaches are still needed for the perfection of AI, and many remarkable inventions and great opportunities await the brave.

References

Baars, B. J. [1997] In the Theater of Consciousness (Oxford University Press).

Baars, B. J. and Franklin, S. [2009] Consciousness is computational: The LIDA model of Global Workspace Theory, IJMC 1(1), 23–32.

Chella A. [2007] Towards Robot Conscious Perception, in A. Chella and R. Manzotti, (ed.) Artificial Consciousness (Imprint Academic) 124–140.

Chella, A. [2008] Perception Loop and Machine Consciousness, APA Newsletter on Philosophy and Computers 8(1), 7–9.

Church, A. [1941] The Calculi of Lambda-Conversion (Princeton University Press).

Dennett, D. [1987] Cognitive wheels: The frame problem of AI, in C. Hookway (ed.), Minds, machines and evolution (Baen Books), 41–42.

Fodor, J. [1975] The Language of Thought (Crowell).

Haikonen, P. O. [1999] An Artificial Cognitive Neural System Based on a Novel Neuron Structure and a Reentrant Modular Architecture with implications to Machine Consciousness, Doctoral Thesis. Series B: Research Reports B4. Helsinki University of Technology, Applied Electronics Laboratory.

Haikonen, P. O. [2003] The Cognitive Approach to Conscious Machines (Imprint Academic).

Haikonen, P. O. [2007] Robot Brains; circuits and systems for conscious machines (John Wiley & Sons).

Haikonen, P. O. [2007b]. Reflections of Consciousness: The Mirror Test. Papers from the AAAI Fall Symposium, Technical Report FS-07-01 pp. 67–71.

Haikonen, P. O. [2011] XCR-1: An Experimental Cognitive Robot Based on an Associative Neural Architecture, Cognitive Computation 3(2), 360–366.

Haikonen, P. O. [2014]. Yes and No: Match/Mismatch Function in Cognitive Robots. Cognitive Computation. Volume 6, Issue 2 (2014), pp. 158–163. doi: 10.1007/s12559-013-9234-z.

Haikonen, P. O. [2019] Consciousness and Robot Sentience Second Edition (World Scientific).

Harnad, S. [1990] The Symbol Grounding Problem. Physica D 42, 335–346.

Hesslow, G. [2002] Conscious thought as simulation of behaviour and perception, Trends in Cognitive Sciences 6(6), 242–247.

Hinton, G. E., McClelland, J. L. and Rumelhart, D. E. [1986] Distributed representations, In D. E. Rumelhart and J. L. McClelland (eds.), Parallel Distributed Processing: Explorations in the Microstructure of Cognition, Vol. 1: Foundations (MIT Press), 77–109.

McCulloch, W. and Pitts, W. [1943]. A logical calculus of the ideas immanent in nervous activity. Bulletin of Mathematical Biophysics, 5, 115–133.

Minsky, M. [1975] A Framework for Representing Knowledge, In P. Winston (ed.). The Psychology of Computer Vision. (McGraw Hill), 211–277.

Newell, A. and Simon, H. [1976] Computer Science as Empirical Inquiry: Symbols and Search. Communications of the ACM 19(3), 902–915.

Rosenblatt, F. [1958] The Perceptron: A Probabilistic Model for Information Storage and Organization in the Brain, Psychological Review 65(6), 386–408.

Samsonovich, A. V. [2010] "Toward a Unified Catalog of Implemented Cognitive Architectures", in A. V. Samsonovich, K. R. Johannsdottir, A. Chella and B. Goertzel (eds.), Biologically Inspired Cognitive Architectures 2010 (IOS Press), 195–244.

Searle J. R. [1984] Minds, Brains & Science (Penguin Books Ltd).

Searle J. R.]1997] The Mystery of Consciousness (Granta Books).

Shanahan, M. [2010] Embodiment and the Inner Life (Oxford University Press).

Steels, L. [2003] Language Re-Entrance and the "Inner Voice", in O. Holland (ed.), Machine Consciousness (Imprint Academic), 173–185.

Taddeo, M. and Floridi, L. [2005]. Solving the symbol grounding problem: a critical review of fifteen years of research. Journal of Experimental & Theoretical Artificial Intelligence.

Zwaan, R. A. and Radvansky, G. A. [1998] Situation Models in Language Comprehension and Memory, Psychological Bulletin 123(2), 162–185.

Chapter 5

Universal Cognitive Intelligence, from Cognitive Consciousness, and Lambda (Λ)

Selmer Bringsjord, Naveen Sundar Govindarajulu, James Oswald

Rensselaer AI & Reasoning (RAIR) Lab,
RPI, Troy NY 12180, USA

We explain that the concept of *universal cognitive intelligence* ($\mathscr{UCI}$) can be derived in part by generalization from the previously introduced (and axiomatized) *theory of cognitive consciousness*, and the framework, Λ, for measuring the degree of such consciousness in an agent at a given time. $\mathscr{UCI}$ (i) covers intelligence that is artificial or natural (or a hybrid thereof) in nature, and intelligence that is not merely Turing-level or less, but also beyond this level; (ii) reflects a psychometric orientation to AI; (iii) withstands a series of objections (including e.g. the opposing position of David Gamez on tests, intelligence, and consciousness, and the complaint that so-called "emotional intelligence" is beyond the reach of any logic-based framework, including thus $\mathscr{UCI}$); and (iv) connects smoothly and symbiotically with important formal hierarchies (e.g., the Polynomial, Arithmetic, and Analytic Hierarchies), while at the same yielding its own new all-encompassing hierarchy of logic machines: $\mathfrak{LM}$. We end with an admission: $\mathscr{UCI}$ by our lights, for reasons previously published, cannot take account of any form of intelligence that genuinely exploits *phenomenal* consciousness.

Contents

List of Figures

1. Introduction; Plan for the Paper

We show herein that the concept of *universal cognitive intelligence* (𝒰𝒞𝒥) can be easily derived in significant part by generalization from the previously introduced and axiomatized *theory of cognitive consciousness* (TCC), and the framework, Λ, for measuring the degree of such consciousness in an agent at a given time (or over an interval of times). Λ (of the finitary variety), in conjunction with TCC and its axiomatization in the system 𝒞𝒜, is presented in (Bringsjord & Govindarajulu 2020). TCC was explained before that publication, more informally and less computationally, in (Bringsjord, Bello & Govindarajulu 2018). Herein, we do not simply reprise this earlier material. Instead, we move from it to 𝒰𝒞𝒥, which applies to all cognitive agents, period. In the case of Λ, we explain efficiently how the new, *in*finitary version of it is very easy to erect.

But what, it will be immediately asked, is cognitive intelligence? We are quite confident that you have considerable cognitive intelligence, and to begin straightaway to characterize 𝒰𝒞𝒥 we take a few minutes here in the introduction to justify our confidence. Our confidence arises in no small part from the fact that performance on certain tests can confirm our attitude, so to a test we now turn.

Imagine that Alice, a math teacher in whose class you are in, asks you to consider the proposition that it's not the case that some proposition ϕ implies another proposition ψ. (Alice is currently teaching basic deductive reasoning in the logic known as the *propositional calculus*, or for us $\mathscr{L}_{PC}$.) Alice's two-part Test 1 for you is that, given this negation (i.e. that $\neg(\phi \rightarrow \psi)$), ...

Test 1

... does it follow deductively that ϕ holds? And part two: Prove that your answer is correct.

We encourage you to take a minute to reflect.

The answer is "Yes," and a valid (and, for exposition here, verbose) proof is easy enough to come by.[1] We assume that you can handle any test question like this at the level of $\mathscr{L}_{PC}$. That is, we assume that you are an agent (and we further assume in particular that you are a *natural*, not an *artificial*, agent) who can decide, for a set Φ of formulae of $\mathscr{L}_{PC}$ whether an arbitrarily given formula ψ also of this logic can be proved from Φ. Where the customary provability relation $\vdash$ is used, and we subscript it with the inference schemata for the particular logic in question, this is the question

---

[1] E.g., this will do:

> **Proof**: Our given is that $\neg(\phi \rightarrow \psi)$. Suppose for indirect proof that $\neg\phi$ holds. Suppose in addition, for a sub-proof using *reductio* as well, that $\neg\psi$; and suppose too — to obtain a conditional directly inconsistent with our given — that ϕ. We deduce ψ by *reductio* from the now-available contradiction, and then since our supposition of ϕ has led to ψ, we deduce the conditional $\phi \rightarrow \psi$. But now we have before us the contradiction of this conditional with our given, and infer by *reductio* on (1) that ϕ. ∎.

Note that only such a proof will do, if we care about scalability, and given the nature of $\mathscr{UCI}$ we very much do. We can't e.g. use a truth table, because such things don't scale up to first-order logic $= \mathscr{L}_1$ and beyond.

(type^2) we assume you can handle:

$$(\dagger) \quad \Phi \vdash_{\mathscr{L}_{PC}} \psi?$$

It may appear to the reader that so far we have said precious little that links cognitive consciousness with intelligence. A skeptic may specifically say Test 1 plausibly relates to cognitive intelligence, but appears to have nothing to do with consciousness. Many readers with little serious exposure to philosophy or psychology in our experience take it for granted that there is a deep nexus between consciousness and intelligence; but those skeptical about our project may claim outright that intelligent behavior in the complete absence of conscious content is entirely possible; and this claim, if left standing, threatens the very foundation of the $\mathscr{UCI}$ conception and framework.

Fortunately (for us, anyway), the claim is already undermined by Test 1; you can see this by returning to this test now, for closer study. Test 1, recall, was presented to you by teacher Alice, whose class, you believe, is intended by her to bring it about in you via your attention-and-perception abilities that you know the answers to instantiations of ($\dagger$). Let the instantiation given above be denoted by ($\dagger^*$). Now, by definition, your level of cognitive consciousness in successfully negotiating Test 1 is present, and indeed quite significant. For, if you are denoted by '$\mathfrak{a}_y$,' and Alice by '$\mathfrak{a}_a$,' and we use for each cognitive verb clearly operative in this simple scenario a modal operator, we can encapsulate the situation by way of the immediately following formula, whose import is likely quite clear to the reader, even without an accompanying tutorial regarding our highly expressive cognitive calculi (to which we return below):

$$(+) \qquad \mathbf{B}(\mathfrak{a}_y, \mathbf{I}(\mathfrak{a}_a, \mathbf{K}(\mathfrak{a}_y, (\dagger^*)))).$$

In essence, this formula says that the test-taking agent (the role played by you, the reader), believes that Alice intends that you know the instantiation of ($\dagger$) holds. This belief on your part intuitively constitutes are significant state of cognitive consciousness. Hence, we see that, albeit quickly and at an

---

[2]The question type here is the "grandfather" of all **NP**-complete problems (though usually the question is put in terms of satisfiability). It thus follows immediately that $\mathscr{UCI}$ includes early on a level of cognitive intelligence in the Polynomial Hierarchy (PH) considered by some to be impressive. For us, as stated later, cognitive intelligence at impressive levels is *beyond* this hierarchy, which accordingly we don't bother to include a view of herein. Readers wishing to initiate study of the Polynomial Hierarchy could turn to (Arora & Barak 2009) for a purely computer-science treatment, and for a truly masterful essay that positions computational complexity/PH within deep issues in mathematics, we heartily recommend (Dean 2019).

intuitive level, that there is in fact an intimate connection between cognitive intelligence, manifested by success on Test 1, and cognitive consciousness.

But what good, you may ask, is the cognitive intelligence we have ascribed to you? Well, if you have this level of cognitive intelligence, you can then use it to find the answer to any problem whatsoever that can be reduced to a query q issued against information Φ (where of course the query is a formula in the propositional calculus). A computing machine with this power, that is to say an *artificial* agent with this level of cognitive intelligence, which consists in automated reasoning, is not a small thing. Historically speaking, the first great success stories for logic-based (logicist) AI, from Simon and Newell, were systems, LOGIC THEORIST and GENERAL PROBLEM SOLVER (GPS), with only a significant portion of this cognitive power; see (Newell & Simon 1956).

Another way to contextualize the level of cognitive intelligence we have introduced via Test 1 is to consider pure "logic machines" and logic programming. The state of research and development of logic machines at the dawn of AI is chronicled in the important (but often overlooked) book *Logic Machines and Diagrams* by Gardner (1958). Put in "Prolog-ish" logic-programming terms, the level of cognitive intelligence we have identified above roughly corresponds to what a Proplog (note the spelling) program can do. Proplog is a proper subset of Prolog in which predicates are strings representing propositions, and terms are absent. Of course, to build artificial agents able to succeed in many areas of AI (albeit to a humble degree) it suffices to create a suitable Proplog program; this can be seen by turning to how the propositional calculus, conjoined with an automated reasoner for it, is shown in action in the AI textbooks of today (see e.g. Russell & Norvig 2020).

Of course, as you and all those working in (or even just seriously interested in) AI will know, the level of cognitive intelligence identified thus far is quite humble. $\mathcal{UCI}$, as is soon seen, scales progressively higher and higher in the levels of cognitive intelligence available, thereby covering agents of increasing cognitive intelligence, *ad infinitum*.

The remainder of the present essay realizes a straightforward plan: We first recapitulate, in brief, TCC, its axiomatization in $\mathcal{CA}$, and (the finitary version of) Λ (§2). We then explain in the context of prior work why a move to universal *cognitive* intelligence is to our minds wise (§3); quickly root $\mathcal{UCI}$ in the psychometric approach to AI introduced and defended by the first author rather long ago (§4); briefly explain how $\mathcal{UCI}$ is naturally associated with Λ of the infinitary form, which is based upon transfinite numbers and infinite matrices (§5); explain that $\mathcal{UCI}$ is connected to the Arithmetic and Analytic Hierarchies, and to the first author's new hierarchy $\mathcal{LM}$ that subsumes these and indeed all standard, established hierarchies (§6); briefly discuss some work by others that relates to $\mathcal{UCI}$ (§7); and present and respond in Sec. 8 to a series of objections to $\mathcal{UCI}$ (including the position of Gamez that consciousness and a test-based approach to intelligence are not compatible, let alone synergistic (§8.3); and the claim that $\mathcal{UCI}$ can't cover forms of intelligence (e.g., so-called "emotional intelligence"; (§8.4)). We wrap up by conceding that non-occurrent mental states are untouched by our framework, and that another form of consciousness, so-called *phenomenal* consciousness, is beyond the reach of $\mathcal{UCI}$ and its ingredients, such as Λ.

2. Cognitive Consciousness, Its Axiomatization ($\mathcal{CA}$), and Λ: Recapitulation

In the present section we briefly recapitulate first TCC, and then Λ.

2.1. *The Theory of Cognitive Consciousness*

As at least philosophers working on consciousness well know, Block (1995) introduced a fundamental dichotomy of types of consciousness: *phenomenal* consciousness (p-consciousness) on the one hand, and *access* consciousness (a-consciousness, abbreviated) on the other. P-consciousness, as most of our readers will know, is "what it's like" consciousness. There is something it feels like to taste, analyze, and enjoy a great bottle of Nerello Mascalese. There is also something it feels like to skydive for the first time. And so on. When you enter these states, you are p-conscious. At least according to the first author, p-consciousness is not only impossible to capture in computation (Bringsjord 1999) — it's not even possible to rationally take a first genuine engineering step toward building an artificial agent that is p-conscious (Bringsjord 2007). We return to p-consciousness at the very end of the present paper.

What about a-consciousness? What is it? We would very much like to provide you with a formal definition, but the concept is, as we've said, due to Block [certainly with antecedents he (1995) discusses], and by any relevant standard he leaves this concept informal.[3] For confirmation to the reader, we convey that Block's (1995) informal definition (p. 231) is as follows: A state of some agent is a-conscious if and only if it is poised (a) to be used as a premise in reasoning, (b) for rational control of action, and (c) for rational control of speech.[4]

Now, a third kind of consciousness is the one near and dear to our hearts: *cognitive consciousness*, or just *c-consciousness*. This brand of consciousness is present only when the agent that bears it has a robust ensemble of cognitive attitudes, which correspond directly to a relevant set of verbs that signal parts of cognition long investigated in cognitive psychology and cognitive science (e.g. see the cognitive verbs that anchor a number of the chapters in the authoritative Ashcraft & Radvansky 2013). The set of these verbs includes: *believing, knowing, perceiving, communicating* (in a natural language, and perhaps also a formal language that might be used in, say, mathematics), *hoping, fearing, intending*, and so on *ad indefinitum*.[5] For the most part, c-conscious states can be denoted by use of gerundive nominals and specifically the schema '**a**'s *V*-ing that ϕ,' where, respectively, these variables take agents, cognitive verbs, and declarative propositions. For us, not only must ϕ be a formula in the formal language $\mathcal{L}$ of some formal logic $\mathcal{L}$, but the schema itself must correspond to some formula in some formal language $\mathcal{L}'$ of some *intensional* logic $\mathcal{L}$. Intensional logics allow for the representation and reasoning over propositions whose meanings are not

---

[3]This is confirmed by the fact that we know not what is meant by 'reasoning' here. For that matter, what is "rational" control of action? In fact, what is rationality? Because of the obscurity of Block's definition, the first author long ago issued a recommendation to discard the term 'a-consciousness' in favor of using instead terms that refer to the kinds of things this umbrella term is supposed to cover (Bringsjord 1997).

[4]Oddly, Block admits (p. 231) that condition (c) isn't necessary, since — as he sees matters — non-linguistic creatures can be a-conscious by virtue of their states satisfying only (a) and (b). This admission seems to us to be indicative of just how murky a-consciousness is — so murky for us that we refrain from addressing such questions as: Is c-conscious content coextensive with Block's admittedly ill-posed definition of a-conscious content? Does a-consciousness play a large role e.g. in certain conceptions of intentional action, keeping in mind that Λ takes account of any intensional operator for 'intends'? We believe that if a-consciousness was defined using the tools and techniques of logic-based AI or CogSci (e.g. see Bringsjord, Giancola & Govindarajulu (forthcoming), Bringsjord 2008), it would be possible to venture formal definitions of c-consciousness. But such an investigation is out of scope for us in the present essay.

[5]As far as we can tell, any agent or system that is cognitively conscious (= i.e. that enters into a series of c-conscious states through an interval of time) is necessarily a-conscious during this stretch. In general, we see no harm in viewing cognitive consciousness to be the most important type of a-consciousness identified by human scientists and engineers thus far.

determined compositionally. For instance, if an agent $\mathfrak{a}$ believes that Selmer is short, and let this be represented by the formula $\mathbf{B}(\mathfrak{a}, S(s))$ (where $\mathbf{B}$ is an intensional operator), whether or not Selmer is in fact short will have no bearing on the truth of this formula. In extensional logics, which include all the elementary classical logics students of logic and mathematics learn when they first start out, for instance the logics $\mathscr{L}_{PC}$ and $\mathscr{L}_1$, if one knows the semantic value of constituents of formulae one can calculate the value of the overarching formula. For instance, if $S(s)$, as an atomic formula in $\mathscr{L}_1$ in which S is a unary relation and s a constant, is true, then we know immediately that the disjunction $S(s) \vee \psi$ is true as well, for any instantiation to ψ.[6] We have introduced an infinite family of logics, *cognitive calculi*, that include intensional operators for all significant cognitive attitudes. For a characterization of cognitive calculi, readers are directed to Appendix A in (Bringsjord, Govindarajulu, Licato & Giancola 2020). For presentation and use of a cognitive calculus that we have made considerable use of in our AI work, the *Deontic Cognitive Event Calculus* ($\mathcal{DCEC}^*$), see (Govindarajulu & Bringsjord 2017, Bringsjord, Govindarajulu & Giancola 2021).

At this point, we recommend that the reader see Fig. 1, which depicts some key aspects of the discussion thus far. The logics along all three "rays" are needed by $\mathscr{UCI}$, and by the new hierarchy $\mathfrak{LM}$ for it. (For those readers wishing to look ahead figure-wise, this new hierarchy is referred to in Figs. 5 and 7, and is generated by the conception of logic programming summarized in Fig. 6.)

2.1.1. *Regarding the Axiom System ($\mathcal{CA}$) for Cognitive Consciousness*

We refrain from covering in the present chapter all the axioms of TCC, since our main purpose is to introduce $\mathscr{UCI}$. For full coverage, the reader interested in more is directed first to the introduction of the axioms (and cognitive consciousness in general) provided in (Bringsjord et al. 2018), and then, for a more detailed (and more technical) presentation of the axioms geared to those in AI, to (Bringsjord & Govindarajulu 2020). It will suffice here to show the reader but two of the simplest axioms of the axiom system $\mathcal{CA}$, and here's the first:

---

[6] For a fuller discussion, but still an economical one, we direct the reader to (Fitting 2015).

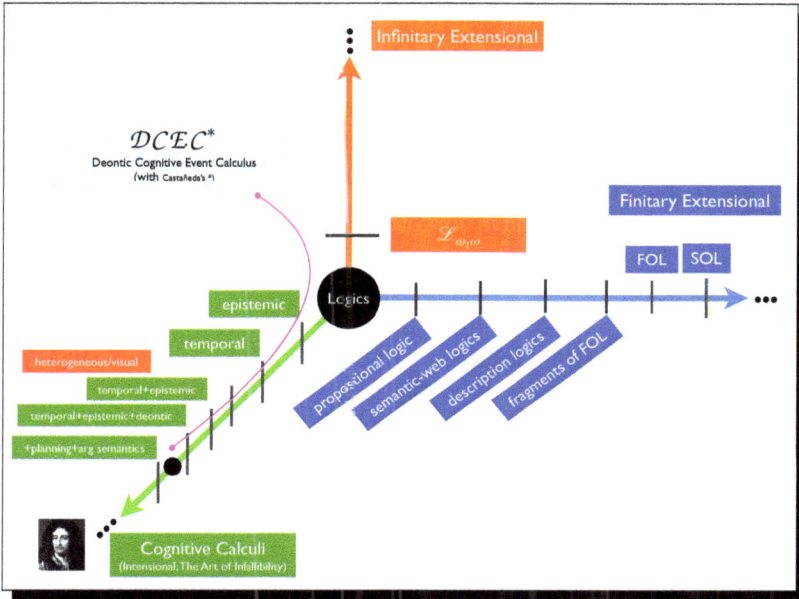

Fig. 1.   The Three "Rays" of Logics. *The blue ray (x axis) emanates out from the start/core, passing through ever-increasingly expressive finitary extensional logics. The orange ray (y-axis) goes upwards, passing through increasingly expressive **infinitary** extensional logics. The green ray is a bit complicated, with details out of scope here, but the basic idea, which has led Bringsjord to proclaim his discovery of Leibniz's "The Art of Infallibility" (in the French used by Leibniz: art d'infaillibilité) is that content from the other rays are combined with full coverage of intensional attitudes in a certain family of intentional logics (called **cognitive calculi**, with diagrammatic/visual logics included as well).*

Perception to Belief

P2B   Human persons perceive internally[a] and externally,[b] and in both cases the percepts in question are believed (at varying degrees of strength, with external perception at the strength of *evident* (which here can be understood as "overwhelmingly likely"), but never *certain*) by these agents, whereas most of what is internally perceived is indeed certain.

---

[a]E.g., we perceive that we are in pain when we are.

[b]E.g., we perceive creatures whose behavior indicates to us that they are in pain.

P2B is easy to comprehend (even without our presenting here the like-lihood calculus and inductive intensional logics that use this calculus).[7] When we perceive such things as that seven is a prime number (via purely mental activity) or that we seem to be sad, we believe these propositions, and they are *certain* for us. But when we perceive in a garden a pink rose, *ceteris paribus* we believe that there is a pink rose before us, but it could be an illusion. (We may have forgotten that we are wearing rose-tinted sunglasses, and we are in fact looking at a white rose.) In c-consciousness as we rigorize it, belief is stratified by its strength (or confidence), in that a belief is accompanied by a *strength factor* σ. The strength of the cognitive attitude of belief is based directly on the underlying *likelihood* of the propo-sition. So for example Jones, if having ingested a powerful pain reliever in a hospital, and knowing that such drugs can have serious side-effects, may believe only at the level of *more probable than not* that there is a walrus before him. With strength stratification in place, belief becomes graded on the positive side from *certain* to *certainly false*, and so will knowledge.[8] (The negative part of the spectrum is symmetrical with the positive. E.g., "more likely than not" $= \sigma$ of 1 has its inverse "less likely than so.") This means that our framework for $\mathcal{CA}$, in contrast with axiom systems based on standard logics (e.g. Peano Arithmetic and axioms for set theory, such as the Zermelo-Fraenkel axioms; these two axiom systems are expressed solely in $\mathscr{L}_1$, which is firmly bivalent), which have binary values TRUE and FALSE, or sometimes those two plus INDETERMINATE, has 13 possible values. In large measure due to the research and engineering of the second author, and significant contributions from Mike Giancola, we have some fairly robust implementations of artificial agents that embody axiom **P2B**, and bring this framework to concrete life; see the simulation, and CPU times, reported in (Bringsjord et al. 2021), and see as well footnote 7, wherein we explain

---

[7]This is a fitting place to explicitly inform the reader that in the present paper we almost exclusively refer to and use *deductive* logics/calculi. We do this to keep things manageable in the span of a single, reasonably sized essay. Were we to enlarge all that we say, all of our figures, and so on, so as to take account of formal inductive logics, the space of a small book would be needed. The best place for diligent readers to turn should they wish to explore how $\mathcal{UCI}$ can be expanded so as to include coverage of the inductive-logic case is (Bringsjord et al. 2021), in which the inductive cognitive calculus $\mathcal{IDCEC}$ is specified and used in a robust simulation with an automated reasoner (ShadowAdjudicator) for it and other such inductive calculi. To our knowledge, this is the only robust automated reasoner for inductive logics at the time the present sentence is being written.

[8]This is what allows us to solve the so-called Gettier Problem. See (Bringsjord et al. 2020).

that the present paper must for economy remain focused, throughout, on deductive logics/reasoning.

Here now is the second axiom from $\mathcal{CA}$ we choose to present:

Introspection (positive)

Intro Humans persons know that they know what they know.

At least the tenor of this axiom is well-known in formal intensional/modal logic because it corresponds to a much-discussed axiom from so-called *alethic* modal logic — an axiom customarily written

$$\Box \, \phi \;\rightarrow\; \Box\Box\phi,$$

when symbolized as the characteristic axiom of the modal logic **S4**, a logic going back to C.I. Lewis. In **S4**, the boxes here are read as 'it's necessary that.' In epistemic logic, we instead read $\Box$ as 'knows that,' denoted by simply '**K**' in our cognitive calculi and in fact in all the intensional logics on the green ray in Fig. 1. A bound $k \in \mathbb{N}$ can be placed on the iteration of **K**, but it would, we think, need to be at least 5 for human-level cognition (for a rationale, see Bringsjord & Ferrucci 2000). The version of the axiom given immediately above has a level of $k = 3$. The axiom here can also be expanded to include provision for negative introspection (i.e., $\neg\mathbf{K}\phi \rightarrow \mathbf{K}\neg\mathbf{K}\phi$), and once again a bound can be placed on the iteration, if desired. Finally, since as we explained in connection with **P2B**, both belief and knowledge have varying strength (or, again, confidence), we have inserted the superscript 'σ' as a variable in the otherwise natural-language version of the axiom **Intro**. In the case of $\mathcal{UCI}$ (and in fact AGI as well; see §7.1), the type of agent a given axiom pertains to doesn't at all have to be those of the human-person variety. But if we stick with **Intro** as we have written it above, and allow '$\mathfrak{a}_{hp}$' to range over human-person agents, and use the epistemic operator **K** for 'knows,' then we have in particular:

$$\forall\phi \; [\mathbf{K}^{\sigma \le 6}(\mathfrak{a}_{hp}, \phi) \;\rightarrow\; \mathbf{K}^3(\mathfrak{a}_{hp}, \mathbf{K}^6(\mathfrak{a}_{hp}, \mathbf{K}^{\sigma \le 6}(\mathfrak{a}_{hp}, \phi)))].$$

It is very important for the reader to understand the implications of the formula given immediately above. Specifically, what we have here implies that in our formal approach to $\mathcal{UCI}$, we must minimally use an intensional logic that allows not only standard quantification over object variables and relation variables (and this to some order in higher-order logic), but must also include formalization of meta-logical propositions. We see this here in that there is quantification over ϕ, which is a variable ranging over all well-formed formula in the logic in question. For ease of exposition and to keep the present essay reasonably sized, we refrain from a full discussion of the fact that cognitive calculi include not only object-level formal languages and inference schemata that regiment reasoning over formulae in these languages, but also meta-logical languages, with formulae in them connected to dedicated inference schemata. That said, the reader can see that proposition (+) above is a case in point, since intensional operators range over meta-logical expressions.

2.2. Λ: *Measuring Cognitive Consciousness*

Λ measures consciousness; specifically, as the reader now knows, c-consciousness; this measurement is based on how the agent in question observably cognizes. Rather than striving to measure phenomenal consciousness (p-consciousness), which is what Φ from Tononi (2012) is supposed to do,[9] Λ explicitly explains and accounts for cognitive consciousness (c-consciousness).

An AI-oriented introduction to Λ is provided in (Bringsjord & Govindarajulu 2020). Later in the present essay, we give a brutally rapid encapsulation of how Λ can be extended to the infinite case, which, as the alert reader by now knows, is certainly needed for the higher reaches of $\mathcal{UCI}$. In the present section we recapitulate finitary Λ; but, relative to Λ defined in (Bringsjord & Govindarajulu 2020), the version of Λ we explain momentarily has been extended, in that now, here, a crucial part of Λ includes the *justification* for the cognition of the agent whose c-consciousness is being measured.

For setting exposition here, assume we have an agent $\mathfrak{a}$ that acts at discrete timepoints. For some of the agent's actions $\alpha(t)$, the agent outputs a justification/rationale *justification*$(\mathfrak{a}, \alpha, t)$. Λ is based on the richness of structures found in the justifications produced by the agent. The justification can be a semi-formal structure, and can include a mix of different modalities (non-verbal actions, gestures, written content, etc.). If the structures include references to cognitive states of other agents or the agent itself, we in general assign a high Λ score to the agent at those points in time. Unlike the aforementioned Φ, we don't provide a single Λ value for an agent or system or creature which is to be measured with respect to c-consciousness; rather, Λ consists in a sequence or vector values

---

[9]For a more technical presentation of the Integrated Information Theory of (p-)consciousness, and Φ, see e.g. (Oizumi, Albantakis & Tononi 2014).

corresponding to the different cognitive verbs discussed earlier as key, such as *knows* **K**, *believes* **B**, *desires* **D**, *intends* **I**, *communicates* or *says* **S**, temporal structures $\vec{t}$, quantifiers $\forall, \exists, \ldots$, etc. Semi-formally, if we have justification *justification*$(\mathfrak{a}, \alpha, t)$ produced by an agent $\mathfrak{a}$ for action α at time t, then (see also Fig. 2 for pictorial exposition):

$$\Lambda\left[justification(a, \alpha, t)\right] = \langle \lambda_{\mathbf{B}}, \lambda_{\mathbf{D}}, \lambda_{\mathbf{I}}, \lambda_{\mathbf{K}}, \lambda_{\vec{t}}, \lambda_{\forall}, \lambda_{\exists} \ldots \rangle$$

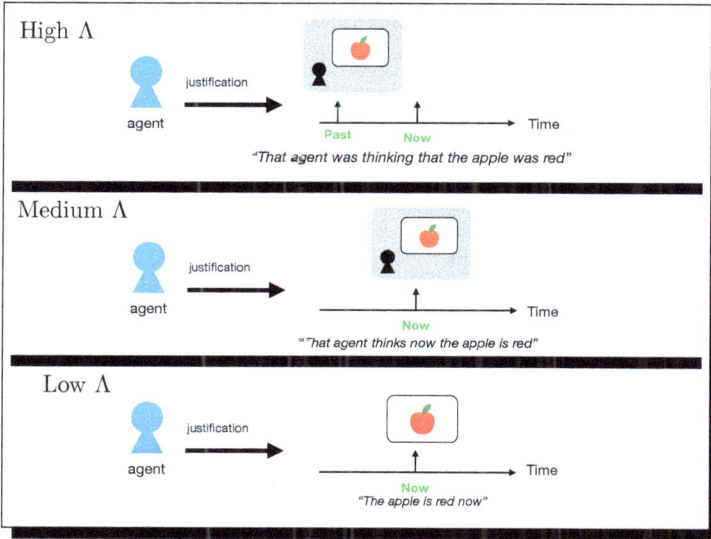

Fig. 2. Λ and Justifications. *We have higher Λ values when the agent has to consider other agents and handle richer temporal structures.*

We end this section with a point about non-modal Λ complexity in formulae and justifications: Because we are herein presupposing the paper in which Λ was introduced (viz. Bringsjord & Govindarajulu 2020), and not fully recapitulating the ins and outs of Λ herein, it may not be evident to some readers of the present essay that differences in the formulae under the scope of modal operators are important determinants with respect to Λ values of formulae. For an example in the present essay, consider the Π_2 formula (2) given later in the paper, quantification in which is somewhat robust. Now consider a simple formula in the propositional calculus, specifically the conditional (denote it by 'c') featured in Test 1 above. Next, we can without loss of generality assume that the reader $\mathfrak{a}_r$ perceives both of these formulae. This means (for us) that both formulae are within the scope of a modal operator $\mathbf{P}$, and there are no other modal operators in play in either case. Yet, clearly Λ applied to $\mathbf{P}(\mathfrak{a}_r(2))$ is significantly greater than Λ applied to $\mathbf{P}(\mathfrak{a}_r c)$. We shall also see in §5.3 that purely extensional elements of formulae inside the scope of modal operators are used to calculate Λ.

3. Why Universal *Cognitive* Intelligence?

An intelligent agent $\mathfrak{a}$ can be artificial. But $\mathfrak{a}$ can also be natural, clearly. After all, since you are currently reading and understanding this sentence (at least up to the right parenthesis that concludes this very remark), you are intelligent — yet you may well not be artificial.[10] You might for instance be a human agent. Therefore, a theory of "universal artificial intelligence" (e.g. that from Hutter 2005) would fail to cover your intelligence (UAI is discussed below, in §7.2). Since we desire to erect a fully comprehensive formal account of cognitive intelligence in agents, whether artificial or natural, finite or infinite, a theory of intelligence that covers only the former is inadequate.[11]

Perhaps, then, we one should seek not a theory of universal *artificial* intelligence, but rather a theory of *computational* intelligence, where 'computational' here is the standard adjective used to cover the processing of information at the level (and below) of a Turing machine and its equivalents.[12] Despite the empirical facts that (i) currently the majority of those working in AI and CogSci restrict their attention to agents capable only of information processing that is indeed merely Turing-level and below (e.g. see the mainstream textbooks, such as Russell & Norvig 2020), and that

---

[10]If God exists, and you are a human agent created by him, then perhaps it would not be inaccurate to say that from his perspective you are an artifact, and hence artificial. But let us leave this possibility aside, and affirm the customary and convenient terrestrial distinction between natural versus artificial agents.

[11]Hutter's (2005) account seems to us inadequate for other reasons. E.g., it fails to insist that no artificial agent can be genuinely intelligent unless it has a large, perhaps even an infinite, amount of declarative knowledge. At a minimum, a cognitively intelligent cognizer must know that it exists, that it has some cognitive states of the formal shape that we have indicated above (e.g., *knowing that it exists*, where here we have instantiation of the schema given above, viz., $\mathfrak{a}$'s V-ing that ϕ, where ϕ is formulae in some formal language for some formal logic/cognitive calculus, and V-ing is cognitive verb in the gerundive-nominal case), etc. As we shall shortly see, intelligent cognizers with high cognitive intelligence have a lot of such knowledge even in the simple case of basic arithmetic. We return to Hutter and his "Universal AI" in §7.2.

[12]Register machines, the λ-calculus, etc. at the level of a Turing machine, and e.g. finite-state automata below this level.

(ii) many who occupy themselves with looking at human brains conceptualize intelligence as constituted and bounded by Turing-level-and-below computation (e.g. Rodriguez & Granger 2016), setting the goal of erecting a comprehensive theory of computational intelligence, with this limited sense of 'computational' affirmed, is an exceedingly bad idea. The reason is obvious: such a theory would fail to cover agents whose information-processing power reaches beyond standard Turing-level machines. For example, an agent able to solve problems that call for *infinite-time* Turing machines (Hamkins & Lewis 2000) would no doubt be rather intelligent, but such an agent stands to a theory of computational intelligence as quantum effects stand to Newtonian mechanics. (As the reader will soon see, $\mathscr{UCI}$ is conceived and designed to be informed by some hierarchies which are mostly about information processing *above* what a Turing machine can do; see, in a look ahead to later in the present paper, Fig. 5. There, both the Arithmetic and Analytic Hierarchies, long established in formal logic/mathematics and theoretical computer science since Turing's dissertation work under Church in the U.S., are mostly super-Turing; and the same holds for the new $\mathscr{LM}$ hierarchy (see Fig. 7).)

What, then, is to be done? We can erect a formal theory of universal *cognitive* intelligence, $\mathscr{UCI}$, one that can cover all cognizers, whether artificial or natural or artificial-natural hybrids, and whether capable of processing information above the reach of standard Turing machines, or not. And how is $\mathscr{UCI}$ to be erected? Well, many steps need to be taken; and many of the steps are (in part) taken herein for the reader to see. But one key step for us was to observe that Λ, which gives a measure of the level of cognitive consciousness of a given agent $\mathfrak{a}$ at any time t, can easily enough be extended to the infinite case. We efficiently explain this extension below, in Sec. 5.3, but first we explain in broad terms another key step on the road to $\mathscr{UCI}$, which is to note that so-called *Psychometric AI* can be extended to allow tests of cognitive power that exceed the Turing Limit on multiple fronts. We explain this step next, in brief.

4. Psychometric AI to Universal Psychometrics

We have earlier introduced a form of AI that is based on the measurement of cognitive ability possessed by given agents. This form of AI is known as, again, *Psychometric AI* (PAI), which was first introduced in (Bringsjord & Schimanski 2003), with subsequent treatment and expansion for example in (Chapin, Szymanski, Bringsjord & Schimanski 2011, Bringsjord 2011,

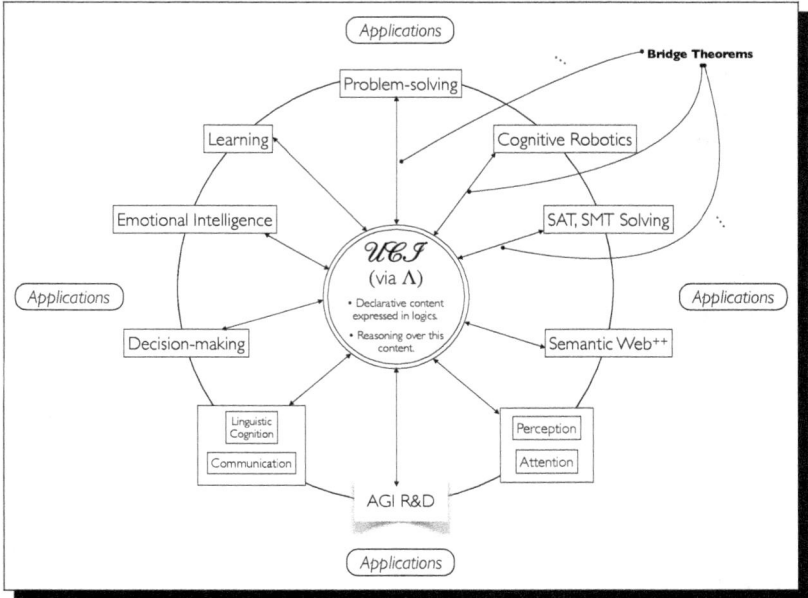

Fig. 3.   Universal Cognitive Intelligence as the Core. *All the areas of AI, AGI, and computational CogSci can be reduced to the logicist formalisms and automated reasoners posited by 𝒰𝒞𝒥, via bridge theorems.*

Bringsjord & Licato 2012, Klenk, Forbus, Tomai & Kim 2011). PAI holds that AI is, or at least ought to be, the field devoted to the conception, design, and implementation of artificial agents able to excel on tests of cognitive ability and skill. Here, we re-affirm the prior commitment to this doctrine — but extend this orientation to agents that are not artificial, not at present terrestrial, not finitary, and not merely Turing-level. For instance, alien creatures that are natural might have exceedingly high levels of cognitive intelligence in connection with arithmetic, as evidenced by their performance on tests of arithmetic. Crucial for understanding this possibility is Fig. 4, which the reader should now consult.

This figure restricts PAI and its tests to be exclusively about arithmetic, but extended to the infinite and super-Turing. All test questions must be restricted to arithmetic involving only the simple arithmetic functions over $\mathbb{N}$: addition, multiplication, subtraction, with relations for $>, <, \geq, \leq$. To put the matter in science-fiction terms, we believe that if alien agents from another planet show up on Earth in a spaceship, they will as a matter of cognitive necessity have had to pass from the innermost circle shown in

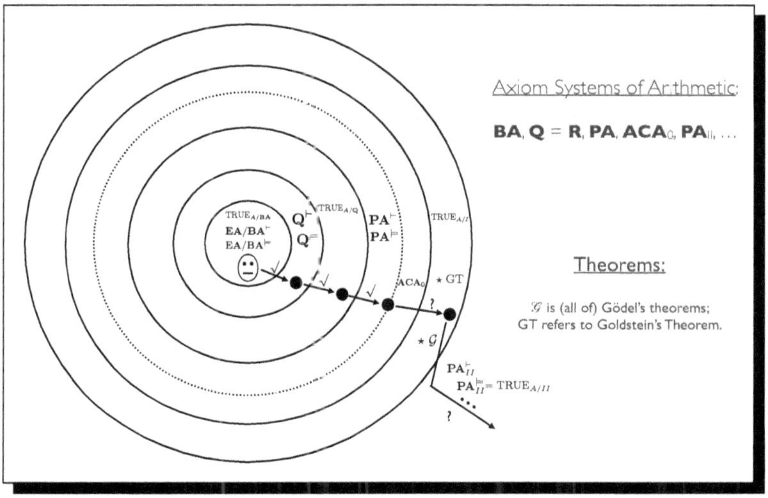

Fig. 4.   The Reverse-Mathematics Basis for Universal Psychometrics (Arithmetic Case). *We assume that all agents with significant cognitive intelligence must "travel" from the innermost circle here outwards, as they master more and more arithmetic. A number of axiom systems for arithmetic are referred to here. **PA** is Peano Arithmetic; this will be familiar to all readers well-versed in formal logic and/or theoretical computer science. Closer to the center of the circles brings one to less powerful systems of arithmetic (for instance "Baby Arithmetic" (**BA**), nicely covered in (Smith 2013)), and "Robinson Arithmetic" (**Q**), nicely covered in (Boolos et al. 2003); farther out from **PA** brings one to more powerful systems of arithmetic (e.g., second-order **PA**). The key thing is the tracking of agents that, as they travel outwards, still understand matters well enough to answer "test questions." For example, and this is a very important feature of the figure, $\mathcal{G}$ denotes Gödel's First Incompleteness Theorem, and a test question that any agent with impressive cognitive intelligence should be able to answer, and justify with a supporting proof, is: "Using the techniques of finitary deduction for $\mathscr{L}_1$, is there in the abstract a proof or disproof of every assertion about arithmetic in the theory of Peano Arithmetic $= \textbf{PA}^{\vdash \mathscr{L}_1}$?"*

Fig. 4 at least out to the minimal circle that includes both $\textbf{PA}^{\vdash}$ (which by default is the deductive closure of the axiom system under first-order inference) and $\textbf{PA}^{\models}$ (closure under standard Tarskian model theory), which is distinguished by an understanding of the theory of arithmetic based upon the axiom system $\textbf{PA}$.[13] In the case of human agents, this first happened, at the latest, by 1930.

---

[13]Excellent introductory coverage of **PA**, sufficient for understanding Fig. 4, is provided in a book the lead author treasures for pedagogy in starting mathematical logic: (Ebbinghaus, Flum & Thomas 1994).

5. Universal Cognitive Intelligence

We now do three things to help further convey to the reader the nature of universal cognitive intelligence, $\mathscr{UCI}$, respectively: We build upon Test 1, given to the reader in our introduction, and issue an additional test (and after that point to further tests), to help the reader better understand at least the early levels in the progression of cognitive power at the heart of $\mathscr{UCI}$ (§5.1); then we venture the start, and *only* the start, of a full, formal definition of an arbitrary level of $\mathscr{UCI}$ (§5.1); and finally, as promised above, we quickly explain how Λ can be extended to the infinite case (§5.3).

5.1. *More Tests to Convey the Idea*

Let us return to teacher Alice from the introduction to the present paper, who now is going to issue Test 2 to you. Alice is actually herself a mathematician, a number theorist, specifically. And here's the background to the new test that she gives you, and we quote what she says to you:

"Recall again the familiar set $\mathbb{N}$ of natural numbers, that is $\{0, 1, 2, \ldots\}$, where of course the ellipsis here indicates that the progression continues infinitely. You have of course long been acquainted with this set; your exposure started in the first grade. Given the natural numbers as our domain of quantification, now consider the arithmetical proposition ν, which I assert to be true, that there is a number $n^\star \in \mathbb{N}$ which is such that, if $n^\star$ is nice, every number is nice. Very well, now your new two-part test question follows immediately."

Test 2
Is Alice's assertion correct? Prove that your answer is correct.

As in the beginning of the present essay, we encourage you to take a minute to reflect, with pen and paper if that is handy for you. ... If you managed to pass Test 2, and you can pass all such tests for arbitrary assertions about the natural numbers, you have cognitive intelligence at a fairly impressive level. This level, in terms of logic-programming, exceeds what Prolog can enable, since the inference schema and process at the heart of Prolog is only guaranteed to succeed when what is to be decided is in fact a theorem. Since $\mathscr{L}_1$ is only semi-Turing-decidable, we would already have to place your level of cognitive intelligence above Turing machines. By the way, now that we have delayed a bit and thereby given you more time, we inform you that Alice is correct — though we leave the discovery of a proof to you, if you don't have one yet.[14] Expressed in $\mathscr{L}_1$, and using obvious symbols for logicizing Alice's claim, we have:

$$\nu : \qquad \exists n[N(n) \rightarrow \forall y N(y)].$$

If we collect the background information available to prove ν into Φ, then, in a direct parallel to what we ascribed to you in the case of Test 1 from the introduction to the present essay, we are here ascribing to you the cognitive intelligence needed to decide:

$$\Phi \vdash_{\mathscr{L}_1} \nu.$$

That level of intelligence, again, can't be achieved by any Prolog program, nor for that matter by any standard automated deductive reasoner operating at the level of $\mathscr{L}_1$.

For a book-length treatment of $\mathscr{UCI}$, we would now continue to give you — in keeping with the avowed psychometric orientation of $\mathscr{UCI}$; recall §4 — tests that see how far up the Arithmetic (which is limited to $\mathscr{L}_1$) and Analytic (which is limited to $\mathscr{L}_2$) Hierarchy you can go, and then we would shift to tests involving intensional operators, such as third-, fourth-, fifth-order ... false-belief tests (Bringsjord, Govindarajulu & C. 2019). In other words, we would be asking you to climb the left-branch progression for Cognitive Calculi shown in Fig. 7.

5.2. *A Full, Formal Definition of $\mathscr{UCI}$?*

Currently, we cannot provide a full, formal definition of $\mathscr{UCI}$. Doing so would mean supplying a formal definiens for at least the following definiendum on the left side of the biconditional:

Def$_\Lambda$ Agent $\mathfrak{a}$ is cognitive-intelligent at level L if and only $\phi(\Lambda)$.

In this definiens, ϕ is itself a formula, one in which 'Λ' occurs, at least once; and L is an integer, real, or infinite ordinal.[15] Obviously it's going to be a challenge to fully flesh out Def$_\Lambda$, in subsequent work and corresponding publication. The present essay serving merely as an introduction to $\mathscr{UCI}$,

---

[14]Hint: Channel the mind of Euclid and approach indirectly, by assuming the opposite of Alice's claim, then use/prove quantifier shift to go from $\neg\exists n \ldots$ to $\forall n\neg \ldots$ in search of your needed contradiction.

[15]Even if we wished to stick with ascending levels of ever-higher cognitive intelligence that climb only extensional logics, we will need to move to transfinite numbers: infinitary logics are *based* upon such numbers; see e.g. (Dickmann 1975), and note the subscript in '$\mathscr{L}_{\omega_1\omega}$,' the one infinitary logic we use in the present essay (§5.3).

the burden of providing a fully instantiated Def_Λ is one we fortunately do not bear at present. But we hope the reader understands the structure of the definition.

5.3. *Expansion of* Λ *for* $\mathcal{UCI}$

Where $\mathfrak{a}$ is, as above, any agent (natural, biological, artificial, supernatural, alien, finite, infinite, divine, etc.) that can consume one or more test questions in the form of a set i of formulae (object-level or meta-formulae), the level of resulting cognitive consciousness is given as Λ, which, as explained in (Bringsjord & Govindarajulu 2020), and reviewed above, is a matrix $\mathbf{X}$. But we have only looked at the finitary case for such matrices before, that is, in (Bringsjord & Govindarajulu 2020). Let's briefly turn now, as promised above, to what the infinitary case looks like:

Example 1

Our agent $\mathfrak{a}$ here knows a basic number-theoretic base formula expressed in the formal infinitary language[16] for the extensional infinitary logic $\mathcal{L}_{\omega_1\omega_1}$ (the smallest infinitary logic), viz.

$$\phi := \exists x_1 \exists x_2 \ldots (x_2 > x_1 \wedge x_3 > x_2 \wedge x_4 > x_3 \wedge \ldots)$$

Assuming that ϕ pertains to the positive integers, this formula can be taken to express in the style of finitary formulae in the axiom system **PA** that integer 2 is greater than integer 1, that integer 3 is greater than integer 2, and so on *ad infinitum*.

Now, what are our Λ measures? Let us have the following four from the scheme of (Bringsjord & Govindarajulu 2020):

μ^1 The "Boolean rank" of a base formula.
μ_i^2 The amount of occurrences of a relation R_i in a base formula.
μ^3 The number of distinct relations in a base formula.
μ^4 The number of quantifiers in a base formula.

Then we have an extension to the infinitary case, unprecedented for Λ relative to earlier work, but simple and easily understood nonetheless:

$$\mathbf{X}_\mathfrak{a} = \begin{pmatrix} \aleph_0 & \aleph_0 & 1 & \aleph_0 \end{pmatrix}$$

---

[16]The language allows infinite disjunctions and conjunctions of the cardinality of $\mathbb{N}$, but restricts the number of quantifiers used in any formula to a particular $k \in \mathbb{N}$. Efficient coverage is provided in (Ebbinghaus et al. 1994).

6. $\mathcal{UCI}$ and the Formal Hierarchies

$\mathcal{UCI}$ harmonizes strikingly well with the standard, formal hierarchies of computational[17] power, including such power above standard Turing machines, because such hierarchies are all based upon ascending complexity and depth of formulae in formal logics.[18] We obviously can't herein systematically canvass these hierarchies, which include the Polynomial, Chomsky, Arithmetic, and Analytic. (As we said earlier in the present essay, we don't include a pictorial overview of the first of these, but such a description of the latter three can, again, be found in Fig. 5.) Of these, which we now proceed to synoptically discuss, we focus upon the Arithmetic, in part because most readers can be counted upon to have an understanding of first-order logic $= \mathcal{L}_1$, which is reasonably well-covered in all the major, comprehensive AI textbooks of today (this holds e.g. for the dominant such textbook: Russell & Norvig 2020).

6.1. *Cognitive Intelligence & the Arithmetic Hierarchy*

In his teaching, the first author has found that it's actually quite easy to build before one's eyes the Arithmetic Hierarchy (AH), and thereby understand it.[19] To do so, one can start with a Turing-decidable relation R to get the climb up AH going. For example, suppose that R logicizes the property/relation of a particular Turing machine $\mathfrak{m}$ taking some particular input a in and, after executing its program, giving some particular b as output in $k \in \mathbb{N}$ steps. Formally:

$$(1) \qquad R(\mathfrak{m}, a, b, k).$$

In terms of cognitive intelligence, you the reader, and indeed all neurologically normal, educated human agents have at least a level of cognitive intelligence sufficient for deciding whether or (1) holds, for any particular quadruple of assignments to the constants the relation is to hold over. (**Proof**: You can simulate the Turing machine in question on the input

---

[17]Those who insist upon reserving 'computation' and 'computational' as terms for Turing-level and sub-Turing-level information processing can simply view the hierarchies in question as describing *information processing*.

[18]Finitary measurement via Λ of the amount of cognitive consciousness in an agent, and as such of the cognitive intelligence in an agent, link directly to these hierarchies, because the formulae that are scored by Λ can be placed within the hierarchies. But we must leave details aside here in the interest of economy.

[19]A conventional, non-do-it-yourself introduction to AH is provided in (Davis, Sigal & Weyuker 1994).

for k steps. ∎) We can also say, from the perspective of Λ, that a human who carries out the cognizing here reaches some particular level of cognitive consciousness = cognitive intelligence. But this level is quite humble. We can climb up in AH, rather quickly, to a level, specifically Π_2, that is much greater cognitive intelligence, and that is above anything a Turing machine can do. A specific case that is a favorite of the first author is the cognitive intelligence needed to create valid computer programs. We assume that in order to do this, the agent in question must be able to judge whether two programs compute the same functions or not (= whether two programs, while syntactically divergent, nonetheless are equivalent). If we assume that this capacity holds in the general case, then using the very same relation R we have just allowed ourselves for (1), we can logicize to obtain this formula:

$$(2) \qquad \forall u \forall v [\exists k [R(\mathfrak{m}_1, u, v, k) \leftrightarrow \exists k' R(\mathfrak{m}_2, u, v, k')].$$

Formula (2), in which of course both $\mathfrak{m}_1$ and $\mathfrak{m}_2$ are free variables, has a leading sequence of universal quantifiers, and — if converted to prenex-normal form so that existential quantifiers are moved to the left — then a sequence of existential quantifiers, and this pattern is what makes it Π_2.[20] An agent able to ascertain whether (2) holds for a pair of Turing machines (or computer programs) has a level of cognitive intelligence at Π_2, and once again we could employ Λ to obtain a score of cognitive consciousness/intelligence as well.

6.2. *The New Logic-Machines Hierarchy* $\mathfrak{LM}$

The "engine" that generates the new logic-machines hierarchy $\mathfrak{LM}$ (see Fig. 7) is *pure general logic programming*, or just PGLP (see Fig. 6), a new logic-programming paradigm for the Bringsjord-found universal rational calculus sought by Leibniz throughout his lifetime.[21] The calculus in

---

[20]We skip some niceties for economy: For example, some readers may ask: "Where's the *set* to be decided? I thought what gets decided (or not) in these hierarchies are sets." The answer is that the free variables give us a set, and we can easily make this explicit by set-builder notation, e.g. by:

$$\{\mathfrak{m}_1, \mathfrak{m}_2 : (2)\}.$$

[21]In a 2016 celebration of Leibniz's death after 200 years, at the University of Turin, Bringsjord announced in his invited lecture that he had found Leibniz's universal rational calculus — and that he would wait to see if anyone could discover his discovery. Later, an in-print declaration was made in (Bringsjord et al. 2021). Many have over centuries pondered and sought what it is that Leibniz sought, and some have proposed some things

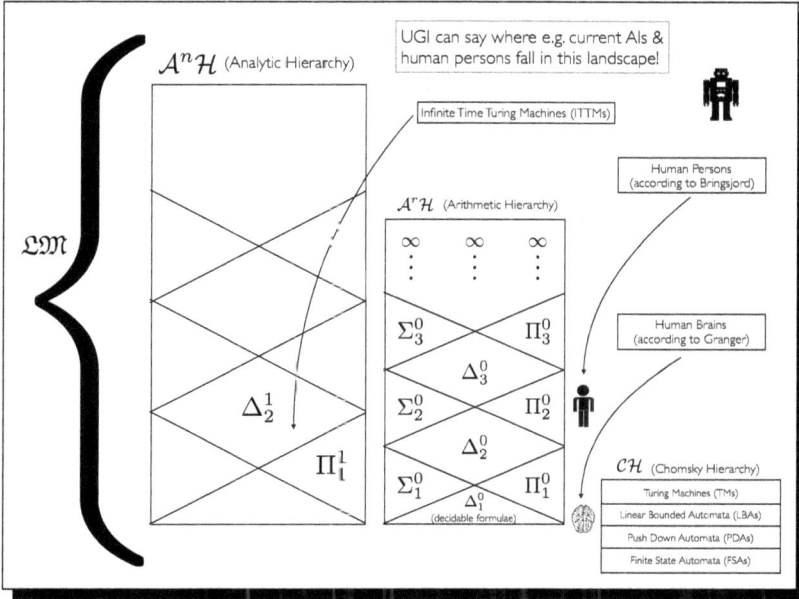

Fig. 5. The Main Logic-based Hierarchies. *The Chomsky Hierarchy will be familiar to most readers; its max is a full Turing machine. The symbols 'Π' and 'Σ' here just indicate the nature of the starting block of quantifiers, either universal in the case of Π or existential in the case of Σ. For coverage of the Arithmetic Hierarchy in purely formal terms, see (Davis et al. 1994), and in connection with AI, see (Bringsjord & Zenzen 2003).*

---

that at least genuinely relate to what Leibniz dreamed of — but they have missed (albeit valiantly) the mark, for many reasons. E.g., consider:

> Today, the best candidates to be considered universal formal languages are the higher-order logics based on type theories, which form the basis of proof assistants such as Coq (Paulin-Mohring, 2015) and Isabelle (Wenzel, 2015). The universality of these logics, from both theoretical and practical points of view, is evidenced by their ability to embed/encode other logics (and even simulate Turing machines) and by their application in many different domains, including Mathematics, Software and Hardware Verification, and even Metaphysics. (Woltzenlogel Paleo 2016, p. 316).

The same position, that higher-order logics at the very least come close to the arrival of Leibniz's dream, is articulated by Benzmüller (2017). Unfortunately, higher-order logics, which form the center trunk/branch in Fig. 7, are subsumed in but absolutely cannot possibly be the universal rational calculus or language Leibniz wanted, for many reasons; here are three: **One**, they are deductive, and as such include no inference question is the language for expressing and defining all cognitive calculi in the family of such, as they have been defined in numerous publications authored by Bringsjord (with others, often), and the framework for *generating* a given cognitive calculus for a given domain of application to be reasoned in and about. PGLP is the *calculus ratiocinator*, the machine or mechanical system that brings the universal rational calculus, or *characteristica universalis*, to concrete, implemented life. But let us dispense with further discussion of such matters, which are best suited for considering Leibniz and his legacy in other venues, and turn to PGLP in earnest, independent of these deeper issues.

PGLP can be viewed abstractly as conforming to the annotated Fig. 6. The three main elements shown in this figure are: $\mathbb{P}$, a program; $\mathbb{R}$, a reasoner; and $\mathbb{C}$, a checker. Territory above the horizontal line is purely specificational;[22] L is simply the background formal language in which both the program $\mathbb{P}$ and query q are expressed. This formal language is bounded only by the bounds of formal logic/meta-logic, whatever they might be. This means, for example, that L might be the object-level formal language underlying the infinitary logic $\mathcal{L}_{\omega_1\omega}$ we used above; in this case, wffs would be allowed to be (countably) infinitely long. The symbols Y, N, and U correspond to "Yes," "No," and "Undetermined." Usually, $\mathbb{P}$ consists simply of a set of formulae against which the query is issued. Programming in PGLP consists in setting this up to issue such a query, and then initiating the activity of the reasoner and the checker. Further details regarding PGLP are beyond the scope of the present essay.

---

schemata in *in*ductive logic, which in its philosophical tradition has informal versions of such schemata (see e.g. (Johnson 2016)), nor do they include provision for probability or likelihood, such as is seen in *pure inductive logic* (Paris & Vencovská 2015); **two**, they are purely symbolic/linguistic, and thus include no inference schemata for diagrammatic/visual information such as are specified in (Arkoudas & Bringsjord 2009); and **three**, they leave aside infinitary logics and infinitary reasoning, since for starters each formulae in higher-order logics is a finite string. The meta-language of cognitive calculi, which is Leibniz's universal rational calculus/language, has this trio of missing elements, and more.

[22]PGLP renders program verification an incalculably easier affair than the well-known burden created by programs in the procedural/imperative, object-oriented, functional, or even standard logic-programming paradigms. The reason is that since specifications *are* programs, program verification reduces to proof/argument checking; this approach was first indicated in (Arkoudas & Bringsjord 2007), and then in the subsequent (Bringsjord 2015) explicitly set out and proposed.

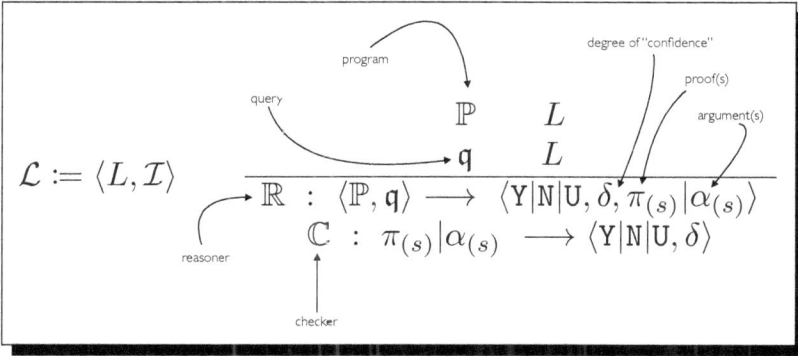

Fig. 6. The Annotated "Engine" for the Hierarchy $\mathfrak{LM}$: PGLP. *Pure General Logic Programming is what generates the hierarchy. In the simple case of PGLP subsuming Proplog and Prolog, the background logic is $\mathscr{L}_{PC}$ and $\mathscr{L}_1$, resp. The hierarchy ascends by virtue of the fact that the power of the combination of logic and corresponding automated reasoner increase.*

6.3. *What About Real-Number-Based Hierarchies?*

Theoretical computer science has traditionally been based upon discrete structures, and the established hierarchies we have visited above are no exception. The reader will have noticed early on that it's the natural numbers which has been a cornerstone herein for most of our presentation of $\mathscr{UCI}$, and of course specifically for Λ (in both the finitary and infinitary cases) and the established hierarchies. There is for example nothing whatsoever "contaminating" AH from the space of uncountable sets, such as the reals $= \mathbb{R}$. What, then, is the relationship between $\mathscr{UCI}$ and the reals (and, indeed, above)? We offer but two remarks, under space limitations. One, there are in fact real-based hierarchies, and we are working out the relationship between $\mathscr{UCI}$ and one of them, one given in (Mycka & Costa 2007). Two, while we have for ease of exposition and focus restricted our attention to $\mathbb{N}$, and to basic arithmetic over this set (recall Fig. 4 and the discussion revolving around it), formal logic can of course be used to capture continuous mathematics, without issue. This we have learned from the sub-part of the discipline of formal logic known as *reverse mathematics*.[23] In general then, universal psychometrics extended to tests regarding such branches of mathematics as analysis, and reasoning over the axiom systems we know to be sufficient to give us continuous mathematics, will be possible in the future to incorporate into $\mathscr{UCI}$.

---

[23] A wonderful starting point for the interested and formally inclined is (Simpson 2010), which covers the reach of various axiom system into mathematics in general.

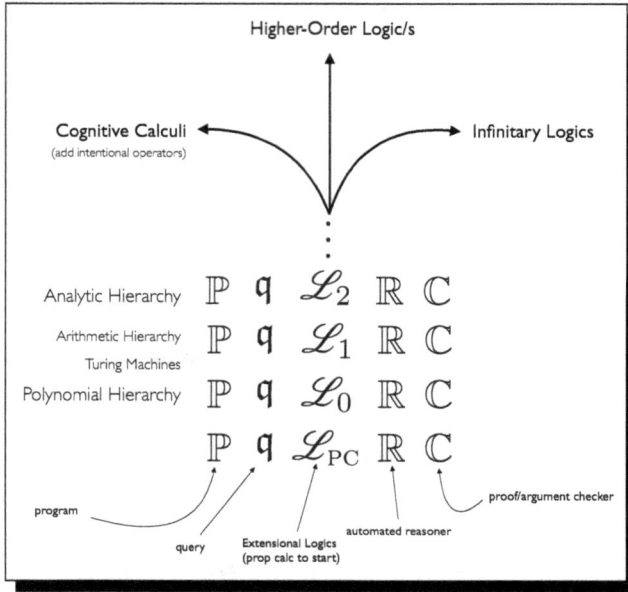

Fig. 7.   An Impressionistic View of $\mathfrak{LM}$ Hierarchy. *When one is referring only to purely extensional, finitary logics, a one-dimensional hierarchy is easy to build, and — at least initially — achieve results with respect to, and then catalogue those results. This would be for the trunk of $\mathfrak{LM}$, and its center branch in the figure here. But there are two other branches, one for intensional logics of ever greater complexity, and one for infinitary extensional logics. The figure here gives merely an impressionistic view because in point of fact many additional branches would need to be specifically shown and charted. For instance, is it clear from the logic given in (Arkoudas & Bringsjord 2009) that any finitary extensional logic can be extended into a heterogeneous logic, that is, a logic that includes not only a linguistic/symbolic formal language, but provision also for diagrams/pictures. Our presentation of $\mathfrak{LM}$ in the present paper suppresses explicit calling out of such nuances.*

7. Related Work: Some Remarks

7.1. *Artificial General Intelligence (AGI)*

AGI, artificial general intelligence, is, at least as some of AGI researchers view their discipline and research, quite relevant to $\mathcal{UCI}$.[24]   To many

---

[24]Outside of AI, we are not aware of related attempts to erect a comprehensive framework along at the general lines of $\mathcal{UCI}$. This is why the present section is limited to discussion of AI under related-to-$\mathcal{UCI}$ work. Of course, one can view the construction and exploration of the standard hierarchies (Polynomial, Arithmetic, and Analytic), and

readers, this is probably plain, to the point that they would fully expect the present section.

A subfield of Artificial Intelligence (AI), Artificial General Intelligence (AGI) can generally be classified as the field that explores the creation of computational agents that possess some level of *general intelligence*: the ability to exhibit complex problem-solving capabilities in an arbitrary environment, akin to the ability of humans (but not necessarily at the same level as humans) (Goertzel 2014, Goertzel 2015, Wang 2019). As AGI focuses on a broad overarching goal, inevitably there are many camps in AGI, each based upon its own approach to the problem (Duch, Oentaryo & Pasquier 2008, Goertzel 2015). These groupings are key to understanding links between $\mathcal{UCI}$ and AGI, as $\mathcal{UCI}$ can be better applied to some groups than to others. Obviously, camps that are not overtly logicist bear little connection to $\mathcal{UCI}$. Here's a simple but nonetheless triadic breakdown of approaches in AGI, the first of which relates most closely to $\mathcal{UCI}$:

- **The Symbolic Approach**: Here logic is in fact the basis for memory and reasoning. Knowledge in these systems consist of statements from which new knowledge can be derived by logical reasoning. New statements may also be added by way of fully logic-based perception see (Wang 2013*b*). Different approaches use different ontologies and different logics with different properties to optimize for the type of reasoning in question (Gust, Krumnack, Schwering & Kühnberger 2009). Invariably, at least so far, relative to the logics upon which $\mathcal{UCI}$ is based, logics in this approach to AGI are inexpressive, and reasoning is correspondingly simple. In particular, anything represented in and reasoned over in this AGI approach can be reduced to information and processing in $\mathcal{UCI}$ (perhaps with tailor-made inference schemata as needed) at the level of only $\mathcal{L}_1$, augmented perhaps with a few intensional operators. Some notable members of the symbolic camp are Wang's NARS (Wang 2013*a*) system and Shaprio's SNePS and GLAIR architectures (Shapiro & Bona 2010), all of which encode symbolic representations of knowledge into a graph representation.
- **The Emergent Approach**: This approach focuses on creating agents whose memory and learning take the form

---

perhaps other non-standard ones (e.g., for quantum computation and real-number-based computation) as quite relevant. (We are not aware of anyone who has done this, but our knowledge is of course limited.) Such a view is, as we have made clear, exactly our own — but at any rate $\mathcal{UCI}$ subsumes all these by virtue of the new $\mathcal{LM}$ hierarchy.

of connectionist systems. The emergent approach assumes, naturally enough, an emergent hypothesis: that symbolic reasoning and learning can emerge from basic connections and interactions between nodes, as they perhaps do (at least in part) in the human brain. "Knowledge" in emergent systems is encoded within the weights and connections between nodes of a network, which may evolve over time for "Learning." $\mathscr{UCI}$ can only subsume emergent approaches indirectly, via reasoning over the declarative content that axiomatizes and thereby captures all connectionist systems. (In this regard, see (Bringsjord 1991).) Direct translation between the sub-symbolic content in such systems to declarative content expressed in one or more logics is impossible, and unwanted. In addition, $\mathscr{UCI}$ would in principle have linkages to processing in artificial neural networks that are more powerful than Turing machines (e.g. analog chaotic neural nets; see Siegelmann 1999).

- **Hybrid Approaches**: Hybrid AGI systems aim to combine emergent and symbolic approaches. According to Duch et al. (2008), hybrid approaches suffer from the same shortcomings as emergent approaches: they have "difficulty in realizing higher-order cognitive functions" such as reasoning over arbitrarily complex/iterated propositional attitudes, which as we have seen are pivotal to cognitive consciousness/TCC. Despite this shortcoming, there is some promise that $\mathscr{UCI}$ could obtain symbolic representations for things at the sub-symbolic level, which would secure the kind of direct connection to $\mathscr{UCI}$ impossible for the emergent approach.

AGI stands in stark contrast to today's mainstream "narrow" AI systems, usually machine-learning models which are trained on massive datasets to excel in particular tasks. For $\mathscr{UCI}$ it is also important to contextualize "human-level" AI in the context of AGI. "Human-level" AI can be thought of as a goal of AGI, but it is only a point on a spectrum of general intelligence that AGI agents fall on. This means that AGI researchers of either a thoroughgoingly or substantive logicist bent can presumably locate their ambitions for future AGI systems in the $\mathscr{UCI}$ space. Unlike other measurements of intelligence, we can quantify very well where humans fall when measured with respect to Λ. Due to the inherently cognitive nature of AGI systems, we think it should be fully feasible, in the future, that any symbolic or hybrid approach in/to AI can be placed within some level/s of $\mathscr{UCI}$. This is indicated by the reference to the relevant bridge theorems between AGI and $\mathscr{UCI}$ in Fig. 3.

One particular point worth nothing is that $\mathcal{UCI}$ stands in contrast to Goertze's conception of intelligence (Goertzel 2021, p. 5), since he writes that "Intelligence in general must be considered as an open-ended phenomenon without any single scalar or vectorial quantification." This of course runs completely counter to the spirit and specifics of $\mathcal{UCI}$. The fact is, in common everyday language we often compare the intelligence of human and nonhuman animal cognitive agents in line with how Λ works. Consider for instance a person "on the street" who states: "Humans are more intelligent than dogs." Clearly intelligence in the sense used here is as a single scalar. However, if our man is asked why he believes humans are more intelligent than dogs, he is likely to resort to informal correlates of measures that are at the heart of Λ. For instance, will our representative human here fail to agree that Fido believes that humans believe that Fido and other canines have no beliefs about the future 10 years hence? Probably not.

7.2. Universal Artificial Intelligence (Hutter)

Marcus Hutter has introduced the concept of "universal artificial intelligence" in his eponymously titled monograph *Universal Artificial Intelligence* (Hutter 2005). His overarching computational model — the *AIXI Model* — features an agent seeking to maximize the reward that it obtains in an unknown environment. The purpose of Hutter's model is to provide a formal, unbounded model for AI. We rest content here with but a few remarks upon Hutter's model. A simpler description than what is provided in the monograph cited immediately above can be found in (Legg & Hutter 2007), and more recently in our own overview of AI (Bringsjord & Govindarajulu 2018) we provide an entire sub-section on AIXI.[25]

Hutter's AIXI model is a formal one of an agent operating in an unknown environment. Given certain reasonable assumptions, this model is universally optimal across all possible environments and has certain provable optimality properties. This model builds upon Solomonoff's Theory of Induction (Solomonoff 1978) and Sequential Decision theory (Sutton & Barto 1998).

More specifically, the AIXI model is an agent-based model in which an agent $\mathfrak{a}$ performs an action in an environment E at time time t. The environment responds with a unique percept and a reward. This continues until some time termed the *horizon* or the *lifetime* of the agent.

---

[25]Go to: https://plato.stanford.edu/entries/artificial-intelligence/aixi.html.

AIXI does have at least one property that in general accords with the fact, presented, explained, and discussed earlier, that $\mathcal{UCI}$, for the most part, is based upon logics whose formulae are allowed to be uncomputable (e.g., infinitely long conjunctions cannot fit on any finite square of a Turing-machine tape), and whose corresponding automated reasoners in $\mathcal{LM}$ are allowed to exceed what a Turing machine can do: AIXI is Turing-uncomputable. At an intuitive level, which is sufficient here, the uncomputability of AIXI is due to the involvement of Turing-uncomputable Kolmogorov complexity and an infinite sum in the specification of AIXI's optimal action. Put intuitively, AIXI is Turing-uncomputable only because it "backs into" Turing-uncomputability. After all, the model rests directly and in no small part on a multi-tape Turing as a fixed part of the theory. In stark contrast, there are levels of cognitive intelligence in $\mathcal{UCI}$ that front and center rest on information-processing that is invariably and essentially beyond the operation of a Turing machine. E.g. consider the cognitive intelligence required to decide Π_2 problems in the Arithmetic Hierarchy, mentioned above. Another example of the fundamental divergence between AIXI and $\mathcal{UCI}$ is clearly seen when one simply takes note of the fact that Kolmogorov complexity, crucial to AIXI, by definition treats *programs* as standard programs only. PGLP programs, in contrast, as we have seen, are arbitrary collections of formulae or meta-formulae constrained only by the formal language of the formal logics or logics selected. In short, and in sum, universal artificial intelligence is certainly not universal cognitive intelligence; and specifically, an artificial agent at the higher infinitary reaches of $\mathcal{UCI}$ far exceeds the cognitive power of any agent in the AIXI framework.

8. Objections; Replies

8.1. *Objection #1: What about intentionality?*

The first objection can be encapsulated as follows: "$\mathcal{UCI}$ inherits from TCC and Λ the centrality of the phenomenon known by philosophers as *intentionality*, a term coined by Brentano.[26] Roughly, as you no doubt well know, intentionality is the 'aboutness' of at least some mental states, especially the very states that you have placed at the heart of $\mathcal{UCI}$. I refer for example to epistemic c-conscious states such as *Selmer's believing that*

---

[26]An overview of intentionality with a level of detail more than sufficient for the present paper is provided in (Jacob 2003/2019).

Naveen believes that Selmer believes that human persons are immaterial.
Here, Selmer, somehow, has a belief that is genuinely *about* Naveen — and
figuring out what it is that makes this the case is an enduring philosophical
conundrum. Moreover, some have claimed that intentionality is itself an
immaterial, or non-physical, phenomenon, and that as a result human persons are *themselves* immaterial. Surely you don't want $\mathcal{UCI}$ to be saddled
with a seemingly insoluble problem, let alone with the long-dead substance
dualism of Descartes."

Our rejoinder is short, simple, and decisive: $\mathcal{UCI}$ will no doubt
raise for many philosophers the spectre of intentionality. But no matter, and no problem; for we are concerned with the abstract *structure* of
c-consciousness, and the all-the-more abstract concept of how to measure
it (via tests, and Λ, of course). We take an intensional approach[27] to formalizing those states that are (for many philosophers of mind) intentional,
but we premeditatedly leave behind the baggage that usually shows up on
the doorstep of such philosophers. Agents blessed with high levels of $\mathcal{UCI}$
enter into robust c-conscious states by definition,[28] and such agents can
reason with and over such states represented in intensional logics, but that
in no ways forces us to affirm even a shred of the positions advocated by
Brentano and followers (such as Roderick Chisholm).

8.2. *Objection #2:* $\mathcal{UCI}$ *is Too Abstract, From the* **Practice** *of AI*

The objection that $\mathcal{UCI}$ is hopelessly abstract has been pressed against us
in more than a few conversations (with antecedents of this objection being
made against us in connection with the TCC/$\mathcal{CA}$/Λ bases of $\mathcal{UCI}$), and
we bring it to the reader's attention here for two reasons: viz., because it
stands to reason that some of our readers, too, will be inclined to object
along this line, and because the objection, while in and of itself anemic, triggers a rebuttal that is important. The rebuttal is expressed, pictorially, in
Fig. 3. This shows our vision that AI of today, AI pursued by *practitioners*,

---

[27]Since we use intensional logics. All our cognitive calculi are intensional logics.

[28]For the record, and to erase any confusion, the reason is simply that if an agent is at
a time or over an interval highly cognitively intelligent, that is because, fundamentally,
put non-technically, the cognitive depth of their states in this period of time is high. But
that that depth is high is precisely what makes their level of cognitive consciousness high.
If for instance an agent provides, with justification, a correct answer to the fourth-order
false-belief task (recall above), they have an impressive iterated belief, and a justification
in support of it, and by Λ that belief and justification will be high — which is to say
that the agent's cognitive consciousness is high.

can (and as far as we're concerned, should) be carried out on the basis of
$\mathscr{U\!C\!I}$. What do we have in mind? The basic idea behind the picture of
Fig. 3 is actually quite simple, as soon as one understands the "bridge" theorems the figure so prominently in it. To understand these theorems, let's
consider one of the sub-areas of AI shown in Fig. 3: "Linguistic Cognition"
and "Communication," each of which have their own box, with both boxes
within a larger one. From a practitioner point of view, this larger box can
be identified with natural-language processing (NLP), which is composed
in AI by natural-language understanding (NLU), and by natural-language
generation (NLG). In the case of NLU, the core challenge is to design and
implement an artificial agent that, taking in natural language, for instance
English, can *understand* that language. Given this, and given the assumption that the type of NLU that stands to be most relevant to $\mathscr{U\!C\!I}$ at least
takes declarative knowledge seriously, it's not difficult to see how bridge
theorems can allow NLU to be reduced to the formal languages, inference
schemata, and automated reasoning that stand at the heart of $\mathscr{U\!C\!I}$. A
great example is the peerless work on knowledge-rich NLU by McShane
& Nirenburg (2021). The reason is that the format in which they represent knowledge, and the processes used to exploit that knowledge in service
of NLU, can all be recast into use of a cognitive calculus that has, as its
finitary extensional component, $\mathscr{L}_1$ and automated reasoning for it.[29]

We would be remiss if we didn't share that our response to the $\mathscr{U\!C\!I}$-
is-too-abstract objection includes that we are actively attempting to build
a physical robot (PERI.2[30]) with unparalleled language-mediated manipulation capacity, and high levels of cognitive consciousness according to Λ,
and correspondingly high levels of cognitive intelligence. This engineering
is in line with robots and consciousness as described in (Chella, Cangelosi,
Metta & Bringsjord 2019), and is designed to take account of the coverage of
artificial consciousness achieved in (Chella & Manzotti 2007). More specifically, and connected to concepts covered in the present essay, in general
conformity with the kind of hybrid architecture presented in (Chella, Frixione & Gaglio 2000), PERI.2 combines logicist representation and reasoning ability (including the treatment of visual information) provided by our
cognitive calculi, but perceptual capability, enabled by ARCADIA (Lovett,

---

[29]Of course, stating and proving the bridge theorem here from frame-based knowledge
representation and reasoning to the formal logics and automated reasoners of $\mathscr{U\!C\!I}$ is
beyond scope, and will have to wait for another day.

[30]The predecessor to PERI.2 was PERI, the robot that concretized Psychometric AI/PAI
discussed herein in §4. PERI and some of its feats are presented in (Bringsjord &
Schimanski 2003).

Bridewell & Bello 2021), that is outside logicist information and processing, but nonetheless tightly integrated with this processing.

8.3. *Objection #3: Gamez*

$\mathcal{UCI}$ runs deeply counter to the claims of Gamez (2020), since, as we have explained at some length, $\mathcal{UCI}$ is (i) based upon the explicit measurement, through Λ, of the complexity of reasoning (which certainly can be counted as a test) on the part of a given agent, and (ii) also overtly upon performance on tests of cognitive ability and skill (as e.g. in the case of testing whether a given agent has deeper and deeper understanding of arithmetic/number theory; recall Fig. 4). For confirmation of the position of Gamez, we for example have this from him:

> In humans we use batteries of tests to indirectly measure intelligence. This approach breaks down when we try to apply it to radically different animals and to the many varieties of artificial intelligence. (Gamez 2020, 51)

As confirmed in this quote, Gamez reasons from two premises to the rejection of approaches like ours that seek to measure arbitrarily high levels of intelligence in machines (i.e. in AIs or artificial agents). We reject both of these premises, as we now briefly explain.

Gamez's first premise is that measuring the intelligence of (nonhuman) animals is problematic. We happily concede this for the sake of argument, despite our knowing that more than a few cognitive scientists have administered cognitive tests to animals, with results, as these researchers see things, confirming that some nonhuman animals are capable of performance on these tests that implies an appreciable level of intelligence (and, indeed, what would in fact be an appreciable level of cognitive intelligence on the scales/spectra of $\mathcal{UCI}$) (e.g. see Taylor, Hunt, Medina & Gray 2008).[31] Nothing follows with regard to $\mathcal{UCI}$ and its formal bases and psychometric roots for the simple reason that it doesn't concern itself with near-zero cognitive intelligence. No nonhuman animals can comprehend domain-independent abstract inference schemata, so for that reason alone they are outside our concern. Recall that we started with Test 1, issued to the reader at the very outset. This test is one the understanding of which requires a prior understanding of domain-independent inference

---

[31] Our view, for what it's worth, is that these investigations are intrinsically inconclusive. We point out that others claim poor performance on the very same class of tasks by monkeys; see e.g. (Visalberghi & Limongelli 1994).

schemata — but we all know very well that such understanding is absent in the case of nonhuman animals. In fact, such creatures have no capacity in natural language at the level of humans (i.e. at the top of the Chomsky Hierarchy), so they can't even read the simplest of the very tests in the 𝒰𝒞ℐ paradigm.

And what of the second premise affirmed by Gamez? Perhaps this one gives rise to an objection that isn't merely a non-starter. This premise is that tests cannot be applied to AIs. But this premise renders any argument such as his circular. What is at issue is whether levels of cognitive consciousness can rationally and otherwise acceptably be mapped to levels of cognitive intelligence to yield 𝒰𝒞ℐ, in symbiotic conjunction with tests of cognitive ability and skill. His second premise is simply the denial of part of the very foundation upon which the 𝒰𝒞ℐ edifice rests. What Gamez needs to provide is some separate, standalone argument as to why cognitive consciousness can't simply be directly, formally connected to assessment via Λ and, in general, tests of cognitive ability and skill along the lines we have presented above (e.g., Is this axiom system for arithmetic consistent?). This will be an acute uphill battle for Gamez, for the simple reason that AI itself, as concretely practiced, places tests front and center in the field, which is why the landmark achievements in AI have so often been constituted by success on tests, usually game-based ones (for chess, checkers, *Jeopardy!*, Go, etc.).

8.4. Objection #4: But Emotional, Artistic, ... Intelligence?

At least some of those skeptical about our logicist approach to cognitive intelligence can, in our experience, be counted upon to object that the emotions, and ergo emotional *intelligence*, is beyond the reach of 𝒰𝒞ℐ. The critic here seems not to have appreciated that according to TCC, cognitive consciousness is reduced to the *formal structure* of cognition in a given agent — and why shouldn't TCC be able to capture such structure in the case of emotions? It is able to do this. Witness that the present objection is exploded in large part by prior work of others making use only of fragments of the highly expressive cognitive calculi we have invented, continue to invent, and avail ourselves of when engineering intelligent machines. A nice example of such prior work is (Adam, Herzig & Longin 2009), in which it is shown that, with the exception of *love*, all the emotions in the dominant account of emotions in cognitive science [the so-called "OCC theory," provided in (Ortony, Clore & Collins 1988)] can be logicized. Additionally, in work along this logicist line, but based upon our cognitive calculi, the second author has led the capture of sophisticated emotions that underlie virtuous behavior; see for example (Govindarajulu, Bringsjord, Ghosh & Sarathy 2019).

9. Two Confessions and Conclusion

Alert readers will doubtless have noticed that nothing above addresses unconscious or generally non-occurrent mental states. Assuming that such states exist in the human case (for our money, they certainly exist in the abstract logico-mathematical space of cognizers in general), the question (Q) arises for us, viz.,

(Q) Can c-conscious states be non-occurrent?

(Q) arises because in fact it's hard to deny that much human activity occurs beneath the surface of awareness, and is subject at least to quasi-rational constraints; consider the simple activities of reaching or walking. In particular for the first author of the present essay, (Q) not only arises, but is pressing, because he has defended at length the proposition that in the human case cognitive intelligence exceeds the Turing Limit $= \Sigma_1$ (Bringsjord & Zenzen 2003),[32] and the information-processing that enables that level of $\mathcal{UCI}$ is thought by him to be at least in significant part unconscious processing. Nonetheless, and here the first confession, we at this point understand c-consciousness, and $\mathcal{UCI}$, to exclusively pertain to occurrent mental states; we thus simply answer (Q) in the negative. Put in terms of problem-solving, and the tests we discussed above (along with many additional tests along the same line), our current assumption about the human case is that unconscious processing only has "intelligence value" insofar as that processing provides to occurrent states and processing outputs that foster answers that are consciously understood and provided as ultimate outputs. Even the first of the tests we presented above, Test 1, may trigger in our readers unconscious processing that yields the answer, but that answer then must be wordsmithed occurrently. Future development of the $\mathcal{UCI}$ paradigm will include deeper consideration and analysis of the unconscious direction.

---

[32]See also (Bringsjord & Arkoudas 2004, Bringsjord, Kellett, Shilliday, Taylor, van Heuveln, Yang, Baumes & Ross 2006).

And now for the second of our confessions: As by now the reader well knows, $\mathcal{UCI}$ is to a significant degree based on TCC and Λ, which, as we've seen, entails that $\mathcal{UCI}$ is based on c-consciousness, which in turn entails that p-consciousness is completely excluded. Putting the matter starkly: A monumentally intelligent agent by $\mathcal{UCI}$, say one that cognizes in the paradise of transfinite levels of Λ courtesy of effortless spinning infinitely long proofs in $\mathcal{L}_{\omega_1\omega}$ in its mind, can at the same time nonetheless be an agent wholly bereft of qualia. This implies that we ought to make a concession, viz. that if there are some types or levels of cognitive intelligence that exploit p-conscious states, $\mathcal{UCI}$ is incomplete. We do not assert *here* that such types or levels exist; but elsewhere one of us has issued such an assertion: (Bringsjord, Noel & Ferrucci 2002). That assertion revolves around what suffices here to be classified as only a concern: the concern, specifically, that *creativity* falls outside of $\mathcal{UCI}$. Future work on $\mathcal{UCI}$ must include systematic investigation of this concern.

Acknowledgements

The invention and introduction of $\mathcal{UCI}$ has been made possible by the generosity of many supporters. In particular, funding from AFOSR (e.g. award #FA9550-17-1-0191) and from ONR (in multiple awards) to Bringsjord and Govindarajulu has been indispensable. An anonymous reviewer insightfully pointed out a number of weak spots in the $\mathcal{UCI}$ conception; we are most thankful, and have done our best to fortify the conception accordingly. Inadequacies that might remain are avowedly our fault. Bringsjord is deeply thankful for being a researcher in the ANR-sponsored PROGRAMme project "What is a (computer) program? Historical and Philosophical Perspectives," participation in which has enabled a number of the ideas presented herein (e.g. PGLP and $\mathcal{LM}$) to be conceived and — at least to a degree — refined. Leadership of PROGRAMme by Liesbeth De Mol and Tomas Petricek is specifically much appreciated. Finally, we are grateful for the ongoing wisdom, patience, and guidance of Antonio Chella, which has been and continues to be crucial to our work, and for his longstanding and continued leadership in the intersection of AI and consciousness.

References

Adam, C., Herzig, A. & Longin, D. (2009), 'A Logical Formalization of the OCC Theory of Emotions', *Synthese* **168**(2), 201–248.

Arkoudas, K. & Bringsjord, S. (2007), 'Computers, Justification, and Mathematical Knowledge', *Minds and Machines* **17**(2), 185–202.
   URL: *http://kryten.mm.rpi edu/ka_sb_proofs_offprint.pdf*
Arkoudas, K. & Bringsjord, S. (2009), 'Vivid: An AI Framework for Heterogeneous Problem Solving', *Artificial Intelligence* **173**(15), 1367–1405.
   URL: *http://kryten.mm.rpi.edu/KA_SB_Vivid_offprint_AIJ.pdf*
Arora, S. & Barak, B. (2009), *Computational Complexity: A Modern Approach*, Cambridge University Press, Cambridge, UK.
Ashcraft, M. & Radvansky, G. (2013), *Cognition*, Pearson, London, UK. This is the 6th edition.
Benzmüller, C. (2017), 'Universal Reasoning, Rational Argumentation and Human-Machine Interaction'.
   URL: *https://arxiv.org/abs/1703.09620*
Block, N. (1995), 'On a Confusion About a Function of Consciousness', *Behavioral and Brain Sciences* **18**, 227–247.
Boolos, G. S., Burgess, J. P. & Jeffrey, R. C. (2003), *Computability and Logic (Fourth Edition)*, Cambridge University Press, Cambridge, UK.
Bringsjord, S. (1991), 'Is the Connectionist-Logicist Clash one of AI's Wonderful Red Herrings?', *Journal of Experimental & Theoretical AI* **3.4**, 319–349.
Bringsjord, S. (1997), 'Consciousness by the Lights of Logic and Common Sense', *Behavioral and Brain Sciences* **20**(1), 227–247.
Bringsjord, S. (1999), 'The Zombie Attack on the Computational Conception of Mind', *Philosophy and Phenomenological Research* **59**(1), 41–69.
Bringsjord, S. (2007), 'Offer: One Billion Dollars for a Conscious Robot. If You're Honest, You Must Decline', *Journal of Consciousness Studies* **14**(7), 28–43.
   URL: *http://kryten.mm.rpi.edu/jcsonebillion2.pdf*
Bringsjord, S. (2008), Declarative/Logic-Based Cognitive Modeling, *in* R. Sun, ed., 'The Handbook of Computational Psychology', Cambridge University Press, Cambridge, UK, pp. 127–169. This URL goes to a preprint only.
   URL: *http://kryten.mm.rpi.edu/sb_lccm_ab-toc_031607.pdf*
Bringsjord, S. (2011), 'Psychometric Artificial Intelligence', *Journal of Experimental and Theoretical Artificial Intelligence* **23**(3), 271–277.
Bringsjord, S. (2015), 'A Vindication of Program Verification', *History and Philosophy of Logic* **36**(3), 262–277. This url goes to a preprint.
   URL: *http://kryten.mm.rpi.edu/SB_progver_selfref_driver_final2_060215.pdf*
Bringsjord, S. & Arkoudas, K. (2004), 'The Modal Argument for Hypercomputing Minds', *Theoretical Computer Science* **317**, 167–190.
Bringsjord, S., Bello, P. & Govindarajulu, N. (2018), Toward Axiomatizing Consciousness, *in* D. Jacquette, ed., 'The Bloomsbury Companion to the Philosophy of Consciousness', Bloomsbury Academic, London, UK, pp. 289–324.
Bringsjord, S. & Ferrucci, D. (2000), *Artificial Intelligence and Literary Creativity: Inside the Mind of Brutus, a Storytelling Machine*, Lawrence Erlbaum, Mahwah, NJ.

Bringsjord, S., Giancola, M. & Govindarajulu, N. S. (forthcoming), Logic-Based Modeling of Cognition, *in* R. Sun, ed., 'The Handbook of Computational Psychology', Cambridge University Press, Cambridge, UK.
URL: *http://kryten.mm.rpi.edu/*
Logic-basedComputationalModelingOfCognition.pdf

Bringsjord, S. & Govindarajulu, N. (2020), 'The Theory of Cognitive Consciousness, and Λ (Lambda)', *Journal of Artificial Intelligence and Consciousness* **7**(1), 155–181. The URL here goes to a preprint of the paper.
URL:
http://kryten.mm.rpi.edu/sb_nsg_lambda_jaic_april_6_2020_3_42_pm_NY.pdf

Bringsjord, S., Govindarajulu, N. & Giancola, M. (2021), 'Automated Argument Adjudication to Solve Ethical Problems in Multi-Agent Environments', *Paladyn, Journal of Behavioral Robotics* **12**, 310–335. The URL here goes to a *rough, uncorrected, truncated* preprint as of 071421.
URL:
http://kryten.mm.rpi.edu/AutomatedArgumentAdjudicationPaladyn071421.pdf

Bringsjord, S. & Govindarajulu, N. S. (2018), Artificial Intelligence, *in* E. Zalta, ed., 'The Stanford Encyclopedia of Philosophy'.
URL: *https://plato.stanford.edu/entries/artificial-intelligence*

Bringsjord, S., Govindarajulu, N. S. & C., E. (2019), Logicist Computational Cognitive Modeling of Infinitary False-Belief Tasks, *in* A. Goel, C. Seifert & C. Freksa, eds, 'Proceedings of the 41st Annual Conference of the Cognitive Science Society', Cognitive Science Society, Montreal, QB, pp. 43–45.

Bringsjord, S., Govindarajulu, N. S., Licato, J. & Giancola, M. (2020), Learning *Ex Nihilo*, *in* 'GCAI 2020. 6th Global Conference on Artificial Intelligence', Vol. 72 of *EPiC Series in Computing*, International Conferences on Logic and Artificial Intelligence at Zhejiang University (ZJULogAI), EasyChair Ltd, Manchester, UK, pp. 1–27.
URL: *https://easychair.org/publications/paper/NzWG*

Bringsjord, S., Kellett, O., Shilliday, A., Taylor, J., van Heuveln, B., Yang, Y., Baumes, J. & Ross, K. (2006), 'A New Gödelian Argument for Hypercomputing Minds Based on the Busy Beaver Problem', *Applied Mathematics and Computation* **176**, 516–530.

Bringsjord, S. & Licato, J. (2012), Psychometric Artificial General Intelligence: The Piaget-MacGuyver Room, *in* P. Wang & B. Goertzel, eds, 'Foundations of Artificial General Intelligence', Atlantis Press, Amsterdam, The Netherlands, pp. 25–47. This url is to a preprint only.
URL: *http://kryten.mm.rpi.edu/Bringsjord_Licato_PAGI_071512.pdf*

Bringsjord, S., Noel, R. & Ferrucci, D. (2002), Why Did Evolution Engineer Consciousness?, *in* J. Fetzer & G. Mulhauser, eds, 'Evolving Consciousness', Benjamin Cummings, San Francisco, CA, pp. 111–138.

Bringsjord, S. & Schimanski, B. (2003), What is Artificial Intelligence? Psychometric AI as an Answer, *in* 'Proceedings of the 18th International Joint Conference on Artificial Intelligence (IJCAI–03)', Morgan Kaufmann, San Francisco, CA, pp. 887–893.
URL: *http://kryten.mm.rpi.edu/scb.bs.pai.ijcai03.pdf*

Bringsjord, S. & Zenzen, M. (2003), *Superminds: People Harness Hypercomputation, and More*, Kluwer Academic Publishers, Dordrecht, The Netherlands.

Chapin, N., Szymanski, B., Bringsjord, S. & Schimanski, B. (2011), 'A Bottom-Up Complement to the Logic-Based Top-Down Approach to the Story Arrangement Test', *Journal of Experimental and Theoretical Artificial Intelligence* **23**(3), 329–341.

Chella, A., Cangelosi, A., Metta, G. & Bringsjord, S., eds (2019), *Consciousness in Humanoid Robots*, Frontiers. Lausanne, Switzerland. ISSN 1664-8714; DOI 10.3389/978-2-88945-866-0.

Chella, A., Frixione, M. & Gaglio, S. (2000), 'Understanding Dynamic Scenes', *Artificial Intelligence* **123**, 89–132.

Chella, A. & Manzotti, R., eds (2007), *Artificial Consciousness*, Imprint Academic, Exeter, UK.

Davis, M., Sigal, R. & Weyuker, E. (1994), *Computability, Complexity, and Languages: Fundamentals of Theoretical Computer Science*, Academic Press, New York, NY. This is the second edition, which added Sigal as a co-author.

Dean, W. (2019), 'Computational Complexity Theory and the Philosophy of Mathematics', *Philosophia Mathematica* **27**(3), 381–439.

Dickmann, M. A. (1975), *Large Infinitary Languages*, North-Holland, Amsterdam, The Netherlands.

Duch, W., Oentaryo, R. J. & Pasquier, M. (2008), Cognitive Architectures: Where do we go from here?, *in* 'AGI', pp. 122–136.

Ebbinghaus, H. D., Flum, J. & Thomas, W. (1994), *Mathematical Logic (second edition)*, Springer-Verlag, New York, NY.

Fitting, M. (2015), Intensional Logic, *in* E. Zalta, ed., 'The Stanford Encyclopedia of Philosophy'.
  URL: *https://plato.stanford.edu/entries/logic-intensional*

Gamez, D. (2020), 'The Relationships Between Intelligence and Consciousness in Natural and Artificial Systems', *Journal of Artificial Intelligence and Consciousness* **7**(1), 51–62.

Gardner, M. (1958), *Logic Machines and Diagrams*, McGraw-Hill, New York, NY.

Goertzel, B. (2014), 'Artificial General Intelligence: Concept, State of the Art, and Future Prospects', *Journal of Artificial General Intelligence* **0**, 1–48.

Goertzel, B. (2015), 'Artificial General Intelligence', *Scholarpedia* **10**(11), 31847. revision #154015.

Goertzel, B. (2021), 'The General Theory of General Intelligence: A Pragmatic Patternist Perspective', *ArXiv* **abs/2103.15100**, 1–64.

Govindarajulu, N. & Bringsjord, S. (2017), On Automating the Doctrine of Double Effect, *in* C. Sierra, ed., 'Proceedings of the Twenty-Sixth International Joint Conference on Artificial Intelligence (IJCAI-17)', International Joint Conferences on Artificial Intelligence, pp. 4722–4730.
  URL: *https://doi.org/10.24963/ijcai.2017/658*

Govindarajulu, N. S., Bringsjord, S., Ghosh, R. & Sarathy, V. (2019), Toward the Engineering of Virtuous Machines, *in* V. Conitzer, G. Hadfield & S. Vallor, eds, 'Proceedings of the 2019 AAAI/ACM Conference on AI, Ethics, and Society (AIES 2019)', ACM, New York, NY, pp. 29–35.

Gust, H., Krumnack, U., Schwering, A. & Kühnberger, K.-U. (2009), 'The Role of Logic in AGI Systems: Towards a Lingua Franca for General Intelligence', pp. 126–131.

Hamkins, J. D. & Lewis, A. (2000), 'Infinite Time Turing Machines', *Journal of Symbolic Logic* **65**(2), 567–604.

Hutter, M. (2005), *Universal Artificial Intelligence: Sequential Decisions Based on Algorithmic Probability*, Springer, New York, NY.

Jacob, P. (2003/2019), Intentionality, *in* E. Zalta, ed., 'The Stanford Encyclopedia of Philosophy'.
  URL: *https://plato.stanford.edu/entries/intentionality*

Johnson, G. (2016), *Argument & Inference: An Introduction to Inductive Logic*, MIT Press, Cambridge, MA.

Klenk, M., Forbus, K., Tomai, E. & Kim, H. (2011), 'Using Analogical Model Formulation with Sketches to Solve Bennett Mechanical Comprehension Test Problems', *Journal of Experimental and Theoretical Artificial Intelligence* **23**(3), 299–327.

Legg, S. & Hutter, M. (2007), 'Universal intelligence: A Definition of Machine Intelligence', *Minds and Machines* **17**(4), 391–444.

Lovett, A., Bridewell, W. & Bello, P. (2021), Selection, Engagement, & Enhancement: A Framework for Modeling Visual Attention, *in* 'Proceedings of the 43rd Annual Conference of the Cognitive Science Society', Cognitive Science Society, Vienna, Austria, pp. 1893–1899.

McShane, M. & Nirenburg, S. (2021), *Linguistics for the Age of AI*, MIT Press, Cambridge, MA.

Mycka, J. & Costa, J. F. (2007), 'A New Conceptual Framework for Analog Computation', *Theoretical Computer Science* **374**, 277–290.

Newell, A. & Simon, H. (1956), 'The Logic Theory Machine: A Complex Information Processing System', *P-868 The RAND Corporation* pp. 25–63. An almost exactly similar version of this paper can be found in *IRE Transactions on Information Theory*, vol **2**, pages 61–79.

Oizumi, M., Albantakis, L. & Tononi, G. (2014), 'From the Phenomenology to the Mechanisms of Consciousness: Integrated Information Theory 3.0', *Computational Biology* **5**(10), 1–25.

Ortony, A., Clore, G. L. & Collins, A. (1988), *The Cognitive Structure of Emotions*, Cambridge University Press, Cambridge, UK.

Paleo, B. W. (2016), Leibniz's Characteristica Universalis and Calculus Ratiocinator Today, *in* C. Tandy, ed., 'Death And Anti-Death, Volume 14: Four Decades After Michael Polanyi, Three Centuries After G. W. Leibniz', Ria University Press, pp. 313–332.

Paris, J. & Vencovská, A. (2015), *Pure Inductive Logic*, Cambridge University Press, Cambridge, UK.

Rodriguez, A. & Granger, R. (2016), 'The Grammar of Mammalian Brain Capacity', *Theoretical Computer Science* **633**, 100–111.

Russell, S. & Norvig, P. (2020), *Artificial Intelligence: A Modern Approach*, Pearson, New York, NY. Fourth edition.

Shapiro, S. & Bona, J. (2010), 'The GLAIR Cognitive Architecture', *International Journal of Machine Consciousness* **02**, 144–152.

Siegelmann, H. T. (1999), *Neural Networks and Analog Computation: Beyond the Turing Limit*, Birkhäuser, Boston, MA.

Simpson, S. (2010), *Subsystems of Second Order Arithmetic*, Cambridge University Press, Cambridge, UK. This is the 2nd edition.

Smith, P. (2013), *An Introduction to Gödel's Theorems*, Cambridge University Press, Cambridge, UK. This is the second edition of the book.

Solomonoff, R. (1978), 'Complexity-based Induction Systems: Comparisons and Convergence Theorems', *IEEE Transactions on Information Theory* **24**(4), 422–432.

Sutton, R. & Barto, A. (1998), *Reinforcement Learning*, MIT Press.

Taylor, A., Hunt, G., Medina, F. & Gray, R. (2008), 'Do New Caledonian Crows Solve Physical Problems Through Causal Reasoning?', *Proceedings. Biological sciences / The Royal Society* **276**, 247–54.

Tononi, G. (2012), *Phi: A Voyage from the Brain to the Soul*, Pantheon, New York, NY.

Visalberghi, E. & Limongelli, L. (1994), 'Lack of Comprehension of Cause-Effect Relations in Tool-Using Capuchin Monkeys (*Cebus apella*)', *Journal of Comparative Psychology* **108**(1), 15–22.

Wang, P. (2013*a*), *Non-Axiomatic Logic, A Model of Intelligent Reasoning*, Sciendo.

Wang, P. (2013*b*), Proceedings of the 6th International Conference on Artificial General Intelligence, *in* 'Artificial General Intelligence', pp. 160–169.

Wang, P. (2019), 'On Defining Artificial Intelligence', *Journal of Artificial General Intelligence* **10**, 1 – 37.

Chapter 6

Intelligence and Consciousness in Natural and Artificial Systems

David Gamez

Department of Computer Science, Middlesex University, London, UK

Humans are highly intelligent, and their brains are associated with rich states of consciousness. We typically assume that animals have different levels of consciousness, and this might be correlated with their intelligence. Very little is known about the relationships between intelligence and consciousness in artificial systems.

Most of our current definitions of intelligence describe human intelligence. They have severe limitations when they are applied to non-human animals and artificial systems. To address this issue, this chapter sets out a new interpretation of intelligence that is based on a system's ability to make accurate predictions. Human intelligence is measured using tests whose results are converted into values of IQ and g-score. This approach does not work well with non-human animals and AIs, so people have been developing universal algorithms that can measure intelligence in any type of system. In this chapter a new universal algorithm for measuring intelligence is described, which is based on a system's ability to make accurate predictions.

Many people agree that consciousness is the stream of colorful moving noisy sensations that starts when we wake up and ceases when we fall into deep sleep. Several mathematical algorithms have been developed to describe the relationship between consciousness and the physical world. If these algorithms can be shown to work on human subjects, then they could be used to measure consciousness in non-human animals and artificial systems.

At present we can use our own imagination, intelligence and consciousness to picture possible relationships between intelligence and consciousness in non-human systems. In the future, we could use mathematical algorithms to measure intelligence, measure consciousness and identify correlations between intelligence and consciousness. This would lead to a more rigorous scientific understanding of the relationships between intelligence and consciousness in natural and artificial systems.

Keywords: intelligence, artificial intelligence, AI, consciousness, artificial consciousness, machine consciousness, prediction, measurement of intelligence, measurement of consciousness, IQ, g-score, universal measure of intelligence, compression, Turing test

1.   Introduction

Humans are highly intelligent, and their brains are associated with rich states of consciousness. We typically assume that animals have different levels of consciousness, and this might be correlated with their intelligence. Very little is known about the relationships between intelligence and consciousness in artificial systems.

Intelligence is a complex multifaceted term and many overlapping definitions have been put forward [Legg and Hutter, 2007a]. Most of these definitions were developed to describe human intelligence and they have severe limitations as definitions of non-human and artificial intelligence.[a] To address this issue, I have developed a new interpretation that links intelligence to a system's ability to generate accurate predictions (see Sec. 2.2).

Many theories have been put forward about the nature of consciousness and its relationship to the physical world. There is less diversity in the definitions of consciousness: many people agree that consciousness is the stream of colorful moving noisy sensations that starts when we wake up and ceases when we fall asleep at night. My version of this definition is given in Sec. 4.1.

These definitions of intelligence and consciousness enable us to carry out preliminary non-scientific work on the relationships between intelligence and consciousness. We can use our knowledge of the domains and our own imagination, intelligence, and consciousness to picture possible relationships between intelligence and consciousness in natural and artificial systems. Some of the insights that can be gained from this approach are covered in Sec. 5.

---

[a]See Burkart *et al.* [2017] for a discussion of intelligence in animals.

A scientific understanding of the relationships between intelligence and consciousness can be developed when we have accurate ways of measuring intelligence and consciousness in natural and artificial systems. Previous work on the measurement of intelligence is covered in the first half of Sec. 3, and Sec. 3.4 describes a new method that I have developed for measuring predictive intelligence. Section 4.3 explains how we can use mathematical theories of consciousness to make deductions about the consciousness of non-human animals and artificial systems. Accurate measurements of intelligence and consciousness will eventually lead to a more systematic understanding of the relationships between intelligence and consciousness in natural and artificial systems.

2. Definitions of Intelligence

2.1. *Previous Definitions of Intelligence*

Intelligence is a complex multifaceted term and many overlapping definitions have been put forward. These include cognitive ability, rational thinking, problem-solving and goal-directed adaptive behavior [Neisser *et al.*, 1996; Bartholomew, 2004; Legg and Hutter, 2007a]. Most of these definitions are based on factors that are linked to intelligence in humans. They often generalize poorly and generate many counterexamples when we try to apply them to non-human animals and to artificial systems.

The problems with defining intelligence have led some people to view intelligence as a collection of abilities. For example, Thurstone [1938] claims that intelligence consists of verbal comprehension, word fluency, number facility, spatial visualization, associative memory, perceptual speed and reasoning. Sternberg [1985] identifies analytical, creative and practical components of intelligence, and Gardner [2006] suggests that there are multiple types of intelligence, including musical intelligence, linguistic intelligence and emotional intelligence. Warwick [2000] frames this more generally with his idea that intelligence is a high-dimensional space of abilities.

A distinction is often made between crystallized and fluid intelligence [Cattell, 1971]. Crystallized intelligence is a stored ability to solve problems. For example, older intelligence tests included factual questions, such as "Who is the president of the USA?". The answers to this type of question must be remembered — they cannot be deduced by reasoning. Crystallized intelligence also includes rules that can be used to solve problems, known as heuristics. For example, it is theoretically possible to deduce how to solve a Rubik's cube from scratch. However, most people use heuristics to solve different parts of the problem — for example, a method for moving a color to a different face — and then sequence the heuristics together to complete the puzzle. Heuristics exist for some of the problems that appear in intelligence tests.

Fluid intelligence is the ability to generalize knowledge and solve problems that have not been seen before. For example, someone with high fluid intelligence might be able to generalize what they have learnt from solving the Rubik's cube to similar puzzles. Modern intelligence tests are mostly designed to measure fluid intelligence. In humans there is a constant interaction between fluid and crystallized intelligence. A solution to a problem might be discovered through fluid intelligence and then stored for rapid recall at a later date.

If we want to understand intelligence in non-human animals and artificial systems, then we need a non-anthropocentric general definition of intelligence that can be applied to any system at all. The next section outlines a number of reasons for thinking that intelligence is closely linked to prediction.

2.2. *Prediction and Intelligence*

Prediction and the Brain

In recent years there has been a surge of interest in the idea that the primary function of the brain is the generation of predictions [Clark, 2016]. According to these theories, each layer in the mammalian cortex[b] generates predictions about activity in the layer below. The layers compare predictions from higher layers with their own activity and pass information

---

[b]Predictive brain theories also apply to animals with different brain architectures, such as cephalopods and birds.

about the prediction errors back up to the layers above. This explains why there are more top-down than bottom-up connections in the brain.

Predictive brain theories typically treat the brain's predictions as probability distributions. This accommodates situations in which we are certain about something, as well as more common scenarios in which we assign probabilities to different events. People working on the Bayesian brain investigate the extent to which the probability distributions of the brain's predictions match the probability distributions of the environment [Knill and Pouget, 2004; Doya *et al.*, 2007].

Bayesian and predictive theories are plausible and attractive interpretations of the brain that are consistent with our subjective experiences. If these hypotheses are partly or wholly true, then the generation of probabilistic predictions is a core function of the brain, and we would expect there to be a strong correlation between a brain's predictive ability and its intelligence.

At the present time there is little direct evidence for predictive interpretations of the brain. Bayesian brain theories are supported by more experiments, but these are controversial – for example, Bowers and Davis [2012] claim that Bayesian models are frequently adjusted to match the data, leading to unfalsifiable theories that are rarely compared with alternatives. There are also ongoing issues with small sample sizes and the reproducibility of experiments in psychology [Collaboration, 2015; Baker, 2016].

A probabilistic and predictive interpretation of intelligence is likely to be attractive to people who already believe that Bayesian and predictive theories of the brain are true. In the future, better evidence could be found for Bayesian and predictive brain theories that would support a link between prediction and intelligence.

Prediction and Action

As I interact with the world, I am constantly predicting the results of different possible actions and selecting the ones that lead to my goals. For example, when I am hungry, I consider the location of different shops and plan how I can get to the best one, considering traffic, petrol, crime, and so on. A system that cannot predict cannot plan — it can only react to changes in its environment as they occur. On the other hand, a system with perfect predictive ability would have God-like omniscience. It would know what would happen under all possible permutations of its environment and could plan sequences of actions that have the highest probability of achieving its goals.

As animals increase in intelligence and complexity there is a shift from hard-wired reactions to planned behaviors based on prediction. Snails follow chemical trails and retreat when danger threatens. The world does something to the snail and it responds in an evolutionarily determined way that, on average, leads to the survival of the species. More sophisticated animals, such as sheep, can classify features of their environment (food, enemies, mates, etc.) and they have a limited ability to predict how their environment will respond to their actions [Gamez, 2019; Marino and Merskin, 2019]. Corvidae (crow family) and cephalopods (for example, octopi and squid), combine reactive behaviors with actions based on richer predictions about their environment, which enables them to solve more complex problems and build tools. Humans combine their reactive behaviors with planning based on complex predictions on multiple time scales.

Prediction and Artificial Intelligence

Systems that are classified as artificially intelligent replicate behaviors that typically require intelligence in humans. Self-driving cars and chess-playing programs are regarded as intelligent because human intelligence is required to drive cars and play chess. Dialysis machines, that replicate the functions of human kidneys, are not regarded as artificially intelligent because blood filtration does not require intelligence in humans.

The problem with this definition of artificial intelligence is that computers can imitate human behaviors in simple ways that do not require intelligence. For example, natural language conversation requires intelligence in humans, but it can be reproduced to a limited extent in chatbots using simple pattern-matching algorithms [Neff and Nagy, 2016]. AI systems that are more plausibly intelligent include game-playing systems, such as AlphaGo, which predict the consequences of different actions in the space of the game. Robots and self-driving cars contain intelligence that enables them to predict the consequences of different actions in the world. AI systems that generate predictions about the future (for instance, climate models) are also regarded as intelligent.

Prediction and Compression

Some people have suggested that intelligence is linked to a system's ability to compress knowledge. This has led to the development of universal measures of intelligence based on compression [Hutter, 2021]. It has been shown that there is a close connection between compression and prediction [Bell *et al.*, 1990], so compression-based theories of intelligence support a link between prediction and intelligence.

Retrodiction/Postdiction

Humans use their intelligence to discover facts about the past as well as the future. For example, historians debate the economic and social consequences of the plague; physicists develop theories about the origins of the universe. This work clearly requires intelligence, and is typically called retrodiction or postdiction.

Spatial 'Prediction'

While prediction is typically thought of as something that occurs across time, we can also make 'predictions' about events that are happening simultaneously at inaccessible points in space. For example, at 4 pm in London, I 'predict' that people are sleeping in Japan. Here I am using my intelligence to reach out beyond the spatial boundaries of my senses.

Predictive Intelligence and Environments

Some people think that intelligence is completely independent of the environment. According to this interpretation, a person has a certain amount of *general* intelligence regardless of whether they are working in the natural world or studying a genomics database.

The most convincing evidence for general intelligence in humans is a correlation between the results of tests of different human abilities, which is usually referred to as *g* [Humphreys, 1979; Haier, 2017]. Experimental work on *g* shows that performance on numerical, spatial, memory and natural language tasks that are *designed for humans* is correlated. However, this research does not show that our ability to perform human-oriented tests generalizes to tasks that are difficult for humans, such as spatial reasoning in high dimensions or identifying patterns in large data sets. The correlation, *g*, only exists because we are comparing tasks that are reasonably easy for humans to complete. No one has demonstrated that human intelligence is general enough to solve all possible types of problem.

Human brains consume a lot of energy and evolution has made many compromises between performance, size, working memory and sensory resolution. Modern human brains are highly capable of solving problems that commonly occur in hunter-gatherer environments, but they have a limited ability to generalize beyond these problems. This point is nicely made by Chollet [2019, pp. 22–23]:

> We argue that human cognition follows strictly the same pattern as human physical capabilities: both emerged as evolutionary solutions to specific problems in specific environments (commonly known as "the four Fs"). Both were, importantly, optimized for adaptability, and as a result they turn out to be applicable for a surprisingly greater range of tasks and environments beyond those that guided their evolution (e.g. piano playing, solving linear algebra problems, or swimming across the Channel) ... Both are multi-dimensional concepts that can be modeled as a hierarchy of broad abilities leading up to a "general" factor at the top. And crucially, both are still ultimately highly specialized (which should be unsurprising given the context of their development): much like human bodies are unfit for the quasi-totality of the universe by volume, human intellect is not adapted for the large majority of conceivable tasks.
>
> This includes obvious categories of problems such as those requiring long-term planning beyond a few years, or requiring large working memory (e.g. multiplying 10-digit numbers). This also includes problems for which our innate cognitive priors are unadapted; for instance, humans can be highly efficient in solving certain NP-hard problems of small size when these problems present cognitive overlap with evolutionarily familiar tasks such as navigation (e.g. the Euclidean Traveling Salesman Problem (TSP) with low point count can be solved by humans near-

optimally in near-linear optimal time, using perceptual strategies), but perform poorly — often no better than random search — for problem instances of very large size or problems with less cognitive overlap with evolutionarily familiar tasks (e.g. certain non-Euclidean problems).

Non-human animals and artificial systems have less generalizable intelligence than humans [Burkart *et al.*, 2017]. Dogs can use their social intelligence to understand humans; they cannot learn to build tools or solve advanced mathematical problems. Birds can adapt nest-building skills to construct wire tools; they cannot learn to play chess. We have built AI systems that can play Go, drive cars, and predict protein folding; no one has developed a completely general AI that can solve any problem or outperform systems built for specific environments.

Humans, non-human animals, and AI systems have, to a greater or lesser extent, limited abilities to solve problems that they have not encountered before. None of the known forms of intelligence are completely general — they can only solve problems within specific environments.

Fluid and Crystallized Predictive Intelligence

When a system encounters a problem or situation that it has encountered before, it can use its memory (crystallized intelligence) to predict what will happen next. For example, the first time that I see a cat crouch down and wiggle its bottom, I might not understand its behavior. Later, when I see the cat pounce on its prey, the meaning of its behavior becomes clear. The next time I see a cat crouching down and wiggling its bottom, I can predict, with reasonable certainty, that it will pounce. I have remembered the sequence of events and can use my memory of this sequence to map earlier events onto later events. This is one form of crystallized predictive intelligence. Crystallized predictive intelligence also includes stored heuristics that we use to solve problems and predict future events.

Fluid predictive intelligence is the ability to make predictions in situations that we have not encountered before. When I first see the cat crouching and wiggling, I might be able to *deduce* from the cat's attitude, preferences, and environment that it is about to pounce. I can do this

D. Gamez

without ever having seen a cat pounce before. In the future I can use my memory to map from the cat's current state to its future behavior — the prediction has moved from fluid predictive intelligence to crystallized predictive intelligence. This transition depends on our ability to *remember* a new prediction or solution so that we can use it more rapidly next time. Without memory our crystallized predictive intelligence remains constant. This is illustrated in the graphs shown in Fig. 1. These graphs suggest that fluid intelligence is linked to an *increase* in our ability to make predictions.

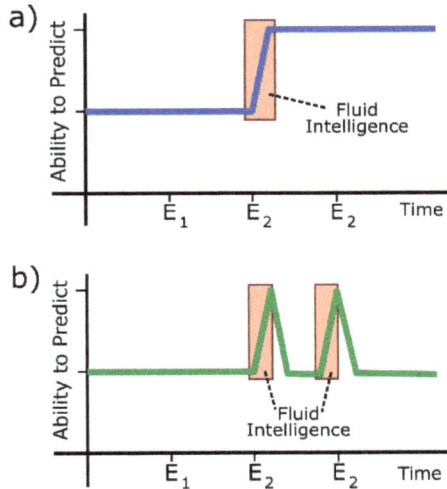

Fig. 1. Illustration of possible relationship between fluid and crystallized predictive intelligence. E_1 is a known event; E_2 has not been experienced by the system before. a) *System with long term memory*. The system predicts the consequences of E_1 using crystallized predictive intelligence. When E_2 first occurs the system uses fluid intelligence to make the predictions. It stores the results, leading to an increase in its crystallized intelligence. The second time E_2 occurs the system uses the stored solution to predict the consequences of E_2. b) *System without long term memory*. When E_2 first occurs it uses fluid intelligence to predict the consequences. This new predictive skill is temporarily held in working memory and is soon forgotten. The next time E_2 occurs it uses fluid intelligence again to make the predictions.

2.3. *Four Hypotheses about Intelligence*

The discussion in the previous section leads to four hypotheses about natural and artificial intelligence:

- **H1**. Prediction is the most important component of intelligence.
- **H2**. Prediction and intelligence are relative to sets of environments.
- **H3**. The amount of a system's crystallized intelligence varies with the number of accurate predictions that it can make in a set of environments.
- **H4**. The amount of a system's fluid intelligence varies with the positive rate of change of its crystallized intelligence.

This interpretation of intelligence fits in well with many previous definitions of intelligence. We can use accurate predictions to plan, achieve goals and receive rewards from the environment. Crystallized predictive intelligence corresponds to our understanding of how things work in the world. However, in this interpretation, knowledge is only linked to intelligence to the extent that it helps us to make accurate predictions. The construction and manipulation of tools also depends on accurate predictions about how manipulations will change the material and how the finished tool will enable us to change the environment. As systems learn, their crystallized predictive intelligence increases. During the time of learning the system will have positive values of fluid predictive intelligence.

When we accept that intelligence is relative to a set of environments (hypothesis H2), it becomes clear that there are many different forms of natural and artificial intelligence, which are specific to their environments. We don't have to broaden our concept of intelligence to handle this (embracing Gardner's multiple intelligences or Warwick's high dimensional space of abilities). Instead, we can say, for example, that system A has a high level of predictive intelligence in a musical environment and system B has a high level of predictive intelligence in a chess environment. These environments can be natural, simulated, data, and so on.

The relativization of intelligence to sets of environments helps us to understand and appreciate the many different forms of human intelligence. IQ tests measure intelligence within an academic environment of mathematical symbols, abstract shapes, etc. This is why IQ tests are correlated with measures of academic success (school grades, advanced degrees, publication of papers, professional careers, etc.). But the academic environment is just one area where human intelligence operates. A successful plumber has a high level of intelligence within the environment of pipes, fittings, water flow, etc. and can make many accurate predictions within this environment. The same is true of other trades and professions. A predictive approach leads to a much broader interpretation of intelligence than the academic intelligence measured by IQ tests.

3.   Measurement of Intelligence

3.1.   *Measurement of Human Intelligence*

Most of the previous work on the measurement of human intelligence has been based on sets of tests that measure behavioral characteristics judged to be linked to intelligence. In the early days these tests included significant numbers of questions based on factual knowledge. Modern human intelligence tests are now mostly based on verbal reasoning, spatial manipulation, and mathematics. The results from these tests are typically converted into values of intelligence quotient (IQ) or g-score. To calculate IQ you take the test results from a sample of the population and calculate the mean and standard deviation. The mean score is assigned an IQ of 100 and each standard deviation above and below the mean corresponds to 15 IQ points. The resulting IQ score can be used to rank individuals according to how well they perform on a battery of intelligence tests. IQ is a population derived measure that does not correspond to a property of a particular individual. Measures of IQ and g-score are controversial and they have often been misused. However, they have played a valuable role in scientific research on intelligence, and they can be an effective way of pre-processing large numbers of applicants for jobs, education, or the military.

Critics of intelligence testing have claimed that intelligence tests only measure the ability of people to complete intelligence tests — they do not actually measure intelligence in the test subjects. In humans this argument is not particularly convincing because human intelligence test scores are correlated with other measures of intelligence. For example, people who score highly in intelligence tests are more likely to achieve advanced educational degrees and pursue careers in areas, such as science, that are generally regarded as requiring intelligence [Robertson *et al.*, 2010; Haier, 2017].

3.2. *Measurement of Intelligence in Non-human Animals*

Animals cannot take human intelligence tests, so there has been a lot of work on the development of cognitive test batteries for animals [Shaw and Schmelz, 2017]. While it might be possible to come up with a plausible set of tests that could be applied to similar animals, this approach is likely to neglect the different types of intelligence that animals develop to survive in their ecological niches. A measure of intelligence that is designed for sheep or fish, for example, cannot easily be transferred to birds or bees. Suppose we want to develop a test that compares human and pigeon intelligence. We could include mathematical abilities and spatial reasoning in our tests, which might be common to both. But pigeons have a greater capacity to map and navigate through their environment, so should this be included in the test as well? As our test battery expands with each species we will end up with a very ad-hoc collection, with each animal scoring well on the tests that are specific to their own set of abilities. It seems highly unlikely that we will be able to design a single set of cognitive tests that would enable us to meaningfully compare intelligence across all species.

A second problem with the measurement of non-human animal intelligence is that we do not have a way of connecting an animal's test results to other indicators of intelligence for that species. Most people would agree that a person who gets top grades in school, gets a first at MIT and publishes ground-breaking physics research is likely to be intelligent. If an intelligence test gives this person a low score, then this is a failure of the test, not an indicator of low intelligence. But how could we ground the results of intelligence tests in octopi, bees, or dogs? Animals do not take advanced degrees or write papers on quantum theory. It is far from clear how we could prove that intelligence tests in animals measure anything more than the ability to perform the test itself.[c]

These problems are often addressed by giving simplified human tests to animals — for example, tests of spatial reasoning or mathematical ability [Boysen and Capaldi, 1992]. These measure the extent to which non-human animals exhibit human intelligence. They are not a meaningful measure of non-human animal intelligence and they do not enable us to compare general intelligence across species.

3.3.   *Measurement of Artificial Intelligence*

Turing testing is often used to measure intelligence in artificial systems. The Turing test was originally proposed by Turing [1950] as a way of answering the question whether a machine could think. He described a thought experiment in which a human and a machine were connected to an electronic typing system and placed in a separate room. A human tester asked the two systems questions and tried to decide which was the human and which was the machine. If the human tester could not reliably identify the machine, then the machine would be judged to be capable of thinking. This test is challenging for machines to pass because the interrogator can ask questions about any topic. Many variants of the Turing test have been proposed. These include embodied Turing tests [Harnad, 1994], behavior in game environments [Hingston, 2009] and the Animal-AI Olympics [Crosby *et al.*, 2019], which provides an environment in which artificial systems can attempt tasks that are believed to require intelligence in animals.

One problem with Turing testing is that as machines improve they are likely to exhaust the possibilities of human tasks. For example, they might eventually map out and completely understand all the possibilities of Go, which would become for them what Tic Tac Toe is for humans — a trivial game whose possibilities can easily be comprehended. To rank AIs according to their intelligence we need tasks that challenge them and which they can complete to different degrees. If they all completely solve

---

[c]These problems are discussed by Legg and Hutter [2007c, p. 5].

a task that is challenging for humans and get the same score, then we can, at most, say that they have super-human intelligence on that task.

A second limitation of Turing testing is that it relies on a clear definition of the human behaviors that require intelligence. In the past it was thought that chess playing was a paradigmatic example of intelligent behavior and that any system that could play chess well would be highly intelligent. Computers can already get an average score on an IQ test [Sanghi and Dowe, 2003] and we now know that low and medium ability chess systems can be built without much intelligence. We are also coming to realize how much of our intelligence is linked to our ability to understand and interact with the natural environment.

Turing testing also cannot measure non-human forms of intelligence. For example, computers are much better at processing vast amounts of data, so they could have much higher levels of intelligence in bioinformatics, while being incapable of solving a Raven's Matrix. It would be extremely anthropocentric to declare that a machine is not intelligent because it cannot solve the narrow range of problems that can be tackled by human intelligence.

To address the limitations of Turing testing, people have developed *universal* measures of intelligence that, in theory, can be applied to any system at all. For example, Legg and Hutter [2007c] developed an algorithm that sums the rewards that an agent receives across all possible environments, with some adjustment for the complexity of different environments. This measure has some intuitive plausibility, but it is not practically calculable because it sums across all possible actions of the agent in all possible environments.

A more practical universal measure of intelligence was put forward by Hernández-Orallo and Dowe [2010], which is based on inductive inference, prediction, compression and randomness. This algorithm is designed to be 'anytime,' which means that whenever it is halted the result should approximate the system's level of intelligence. One issue with Hernández-Orallo and Dowe's approach is that they test the agent in balanced environments in which random actions lead, on average, to zero reward. In practice this is unrealistic because most environments in which intelligence operates are not balanced. If we try to sum intelligence across many unbalanced environments, then the intelligence could be drowned out by random noise. The time limitation of their anytime measure also fails to capture the intelligence of systems that operate on long time scales, such as trees and human societies.

Other people have developed universal measures of intelligence based on compression. In the Hutter prize people compete to compress 1 GB of Wikipedia data [Hutter, 2021]. Hernández-Orallo's C-test measures the ability of a system to find the best explanation for sequences of increasing complexity in a fixed time [Hernández-Orallo, 2000]. The best explanation is usually a compressed version of the sequences that enables the subject to predict more sequences of the same type. As explained in Sec. 2.2, good compressors are good at prediction, so these kinds of tests go some way towards measuring predictive intelligence. However, predictive intelligences are typically specialized for the different areas that the operate in (motor movement, psychology, stock prices, protein folding, etc.), so a single type of compression test only provide one way of estimating predictive intelligence in a limited class of systems.

More detailed summaries of some of these measures of artificial intelligence are given by Legg and Hutter [2007b]. Hernández-Orallo [2017] discusses other ways in which the performance of AI systems can be evaluated.

3.4. $\cancel{K}$: *A New Universal Measure of Predictive Intelligence*

This section describes a new universal algorithm that I have developed to measure predictive intelligence. This algorithm works by comparing the probability distributions of a system's predictions with the probability distributions that actually occur (see Fig. 2).

Prediction Accuracy

To measure the accuracy of a system's predictions we need to compare the probability distributions of the predictions with the later probability distributions of the internal states. For example, in Fig. 2, we compare prediction P_{1-2} made at time 0 with I_{1-2} at time 2. For discrete probability distributions the prediction accuracy is measured using Hellinger distance:

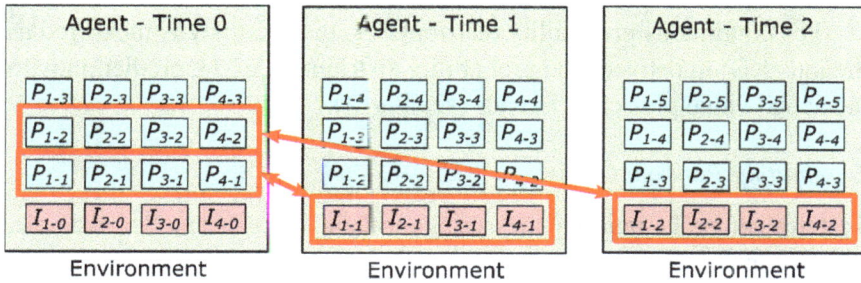

Fig. 2. Agent's predictions about its internal states. The agent has internal states I_1, I_2, I_3 and I_4. I_{1-0}, I_{2-0}, I_{3-0} and I_{4-0} are the probability distributions of the internal states at time 0. P_{1-1}, P_{1-2}, P_{1-3} are predictions that the agent makes about the values of I_1 at times 1, 2 and 3. As the spatial and temporal properties of the environment change, the future states of I_1, I_2, I_3 and I_4 are compared with earlier predictions to evaluate their accuracy.

$$H(P,Q) = \frac{1}{\sqrt{2}}\sqrt{\sum_{i=1}^{k}\left(\sqrt{p_i} - \sqrt{q_i}\right)^2} \tag{1}$$

Hellinger distance is 0 when there is an exact match between two probability distributions, and 1 when there is a complete mismatch. So 1-$H(P,Q)$ gives the *degree of match* between two discrete probability distributions, P and Q, expressed as a number between 0 and 1.

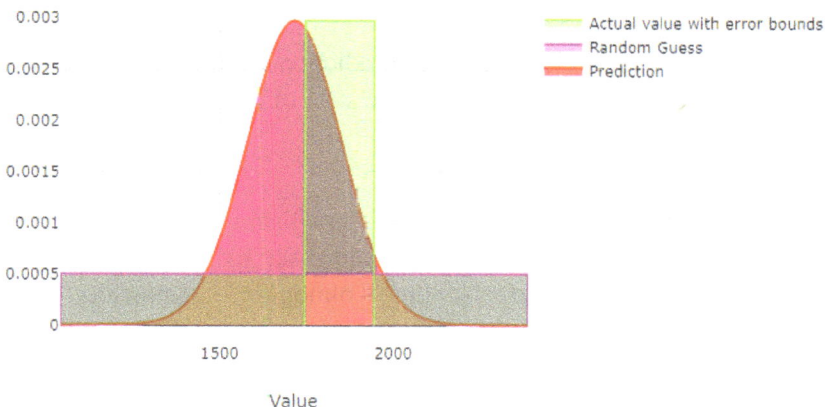

Fig. 3. Prediction match for continuous probability distributions. Here the prediction is a normal distribution. The actual value is discrete and error bounds are added so that there is a non-zero intersection with the prediction. The random guess is an equal distribution across the known range of values.

For continuous probability distributions, the accuracy is measured as the intersection between the probability distribution of the prediction with the actual value plus error bounds (see Fig. 3).

Eliminating Random Guesses

Systems can make random guesses about their future states. For example, suppose that I_1 in Fig. 2 can have discrete values 1, 2 or 3. In this case, a random guess would be $p(I_1 = 1) = 0.3333$, $p(I_1 = 2) = 0.3333$ and $p(I_1 = 3) = 0.3333$. This guess does not require a significant amount of intelligence because it is generated without any data from the environment. However, it will still have some match with the states that actually occur at the next point in time. In discrete probability distributions this issue is addressed by subtracting the random guesses from the prediction match, as shown in Eq. (2).

$$PM(P, I) = \left| \left(1 - H(P, I) \right) - \left(1 - H(R, I) \right) \right| \tag{2}$$

PM is the match between a prediction, *P*, and an internal sensory state *I*. *R* is an equal distribution across possible sensor values. With continuous probability distributions, the random guess is interpreted as an equal distribution across the range of previously seen values, as illustrated in Fig. 3. In both cases the absolute value of the result is taken to prevent negative values of intelligence. The prediction match is summed for all predictions that are made at each unique state of the environment.

Trivial and Non-trivial Predictions

Systems could artificially increase their intelligence by generating large numbers of trivial predictions. This is addressed by multiplying the sum of the prediction matches by the Kolmogorov complexity of the predictions, as shown in Eq. (3):

$$PM_e = \frac{K(l)}{L(l)} \sum_{s=1}^{p} \sum_{i=1}^{q} \sum_{t=1}^{r} PM(P_{i-t}, I_{i-t}) \tag{3}$$

PM_e is the total prediction match for a single environment, *l* is a string that describes all the predictions in the environment, $K(l)$ is the Kolmogorov complexity of *l*, and $L(l)$ is the length of *l*. The predictions are summed for all unique states of the environment ($s = 1 ... s = p$), for all internal states

($i = 1\ldots i = q$) and for all times covered by the predictions ($t = 1\ldots t = r$). Kolmogorov complexity cannot be directly calculated, so it is often approximated by compression algorithms.

Comparing Simple and Complex Systems

To make it easier to compare systems with large differences in intelligence, the log is taken of the total prediction match for the environment (PM_e). This makes it easier to compare highly complex systems, such as humans, with trivial AI systems on the same scale. In some cases PM_e will be less than 1. This corresponds to a low level of intelligence, but the log of a very small number is a large negative number, which makes no sense for intelligence. So the log is only taken when PM_e is greater than 1. This leads to the final equation for calculating the predictive intelligence, PI_e, of an agent in environment e:

$$PI_e = \begin{cases} log_2(PM_e) & if\ PM_e > 1 \\ 0 & otherwise \end{cases} \qquad (4)$$

Summing Intelligence across Environments

Predictive intelligence is relative to a *set* of environments (hypothesis H2), so we need to sum PI_e for all the environments in the set. This raises the problem that we don't have a clear definition of what constitutes a distinct environment. Some environments are very different from each other; others only have trivial differences between them. If we simply add the PI_e values from different environments together, an agent will double its intelligence across two environments that are almost identical. To address this issue, the sum of PI across environments $E_1, E_2, \ldots E_n$ is multiplied by the Kolmogorov complexity of the combined environments divided by the sum of the Kolmogorov complexity of the environments considered independently:

$$K_c^{1.0} = \frac{K(E_1 + E_2 + \cdots + E_n)}{K(E_1) + K(E_2) + \cdots + K(E_n)} \sum_{e=1}^{p} PI_e \qquad (5)$$

Environments appear very differently to systems with unique senses and diverse ways of processing sensory data. So the complexity of environments has to be considered from the perspective of the agents, not from some fictional standpoint of complete objectivity. If two environments are very similar, then the joint complexity will be approximately the same as the individual complexity and the first half of Eq. (5) will be approximately 1/2. On the other hand, if two environments are very different, then the shortest program describing both will be approximately equal to the sum of the lengths of the shortest programs describing the environments individually. In this case the first half of Eq. (5) will be approximately 1. In practice, Kolmogorov complexity is calculated using compression algorithms, which need to be carefully chosen to take the nature of the environments into account.

The letter that I have chosen to represent this measure of intelligence is $\math{K}$, which is the Old Norse letter (rune) that corresponds to our modern 'p' sound. $\math{K}$ is pronounced 'peorth', 'perth' or 'pertho'. The $\math{K}$ rune is associated with the dice cup, chance, secrets, destiny, and the future, which is appropriate for a measure that is based on a system's ability to make predictions. In Eq. (5), the subscript, c, indicates that it is crystallized predictive intelligence. The superscript is the version of the algorithm.

Fluid Intelligence

Fluid intelligence corresponds to a system's ability to learn and adapt to its environment. It is linked to positive changes in a system's crystallized intelligence (see hypothesis H4), as shown in Eq. (6):

$$\math{K}_f^{1.0} = \begin{cases} \dfrac{d\math{K}_c^{1.0}}{dt} & if \ \dfrac{d\math{K}_c^{1.0}}{dt} > 0 \\ 0 & otherwise \end{cases} \tag{6}$$

Experiments

The feasibility and performance of the algorithm have been tested on an agent in a variety of maze environments, and on a deep neural network that performs time series prediction. The experiments are implemented as a website: www.davidgamez.eu/pi, which is shown in Fig. 4.

These experiments show that $\math{K}$ is straightforward to measure on artificial systems when we have full access to their internal states and the environments can be fully explored. More work is required to develop ways of estimating the predictive intelligence of less accessible artificial systems that partially explore their environments.

a) ꝓ: A Universal Measure of Intelligence based on Prediction

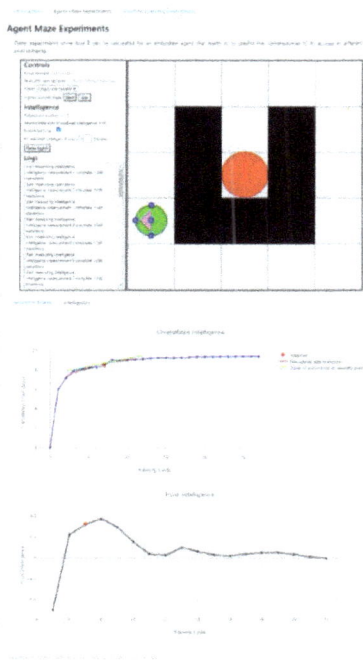

b) ꝓ: A Universal Measure of Intelligence based on Prediction

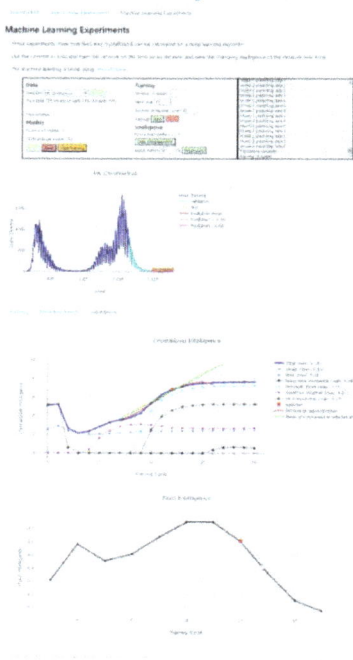

Fig. 4. Website with experiments that test ꝓ algorithm. a) Fluid and crystallized ꝓ are calculated for an agent that predicts the consequences of its actions in different maze environments. b) Fluid and crystallized ꝓ are calculated for a deep network that predicts future values in different time series (synthetic, stock prices, weather, and coronavirus).

We have very limited access to natural systems' internal states. Brain activity can be read non-invasively using fMRI, MEG, and EEG, but these technologies have very low spatial and/or temporal resolution. Optogenetics can give us close to real time measurements of the entire brain of small transparent organisms, such as Zebrafish larvae [Portugues *et al.*, 2013], and it might be possible to apply this to other transparent animals, such as the glass octopus. With non-transparent animals we can only measure ~20,000 neurons on the surface of the brain in real time with current technology. Ways will have to be found to estimate the predictive intelligence of these systems from limited data and from external behavior. We will then be able to use ꝓ to systematically study and compare the intelligence of humans, non-human animals, and artificial systems.

4.   Consciousness

4.1.   *Definition of Consciousness*

When we are conscious we are immersed in a bubble of space, roughly centered on our bodies, within which objects and non-physical properties, such as color and smell, are distributed. I describe this as a *bubble of experience* [Gamez, 2018]. My bubble of experience currently contains green trees and the smell of coffee. When I am at the beach my bubble of experience contains white sand, blue sea, and the taste of tequila. In online perception objects and properties in our bubbles of experience co-vary with the physical world. We can also change our conscious experiences offline, independently of the world, in dreams and imagination.

Bubbles of experience have multiple dimensions of variation. The spatial size of bubbles of experience can vary, there is variation in temporal depth [Husserl, 1964] and there can be more or less objects and properties and more or less types of objects and properties. The contents of bubbles of experience can also appear with different levels of intensity. In dreams, imagination and on the edge of sleep, contents are vague, washed out and unstable. In online perception contents are vivid and stable with rich colors. A person on hallucinogens can have experiences with greater intensity than the normal waking state. The contents of a single experience can have range of intensities. There might be a fleeting impression of a bird at the edge of my field of vision while I am looking at a bright red bus rushing towards me and experiencing intense feelings of fear and panic.

There are challenging philosophical problems with consciousness, such as the hard problem and the relationship between consciousness and the physical world. Elsewhere I have shown how our modern concept of consciousness (and some of its problems) co-evolved with the development of modern scientific theories about the physical world [Gamez, 2018].

4.2. *Physical, Computational, Functional and Informational Theories of Consciousness*

Physical theories of consciousness link consciousness to spatiotemporal patterns in particular physical materials. For example, there are neural theories of consciousness [Koch *et al.*, 2016], electromagnetic theories of consciousness [Pocket, 2000] and quantum theories of consciousness [Hameroff and Penrose, 1996]. Physical theories of consciousness are similar to other scientific theories that are based on spatiotemporal physical patterns: moving electrons produce magnetic fields; moving neutrons do not.

Many people believe that consciousness is linked to computations or functions [Cleeremans, 2005]. They claim that consciousness is present wherever a particular computation or function is executed, independently of how the computation or function is implemented. For example, people have connected consciousness with the implementation of a global workspace [Dehaene, 2014]. Information integration theory connects patterns of information to consciousness, independently of the physical implementation of the information [Tononi, 2008].

Physical, computational, and functional theories of consciousness have some common ground. It might be the case that global workspace theory, for example, captures a pattern, which is linked to consciousness when it is implemented in a biological brain. However, computational and functional theories of consciousness lose plausibility when the claim is made that a computation or function is linked to consciousness *independently* of the material in which the computation or function is realized. One problem with this claim is that a system executing a computer program is a sequence of physical states, and Putnam [1988] and Bishop [2009] show that any sequence of physical states can be interpreted as implementing a particular run of a given program. This leads to an implausible panpsychism and to the untenable result that every brain is associated with an infinite number of different consciousnesses.

A second problem with computational and functional theories of consciousness is that they can only be scientifically tested if we have an objective way of measuring the presence or absence of a computation or function in a system. For example, to prove that global workspace theory is correct, we need to be able to determine whether there is an active global workspace in the conscious brain and show that no global workspaces are being executed in the unconscious brain. Unfortunately we do not have a way of unambiguously measuring the computations or functions that are being executed in a physical system [Gamez, 2014a]. Information integration theory has similar problems with the subjectivity of information and with the measurement of information in a system [Gamez, 2016]. The only reasonable conclusion is that computations, functions, and information are subjective — not objectively measurable properties of physical systems. Consciousness must be linked to objective physical properties of a system.

4.3. *Measurement of Consciousness*

To study the relationships between consciousness and the physical world we need to measure consciousness, measure the physical world and look for connections between these sets of measurements.

Consciousness is measured through first-person reports. For example, when I am eating an apple, I can describe its red color and sweet flavor. We believe our own first-person reports and typically believe the first-person reports of other adult humans (disregarding philosophical problems with zombies, etc.), because other people have similar brains and we assume that there is a close relationship between physical and conscious states.

Infants, animals, and computer programs also generate first-person reports about consciousness. However, these systems have different brains or no brains at all, so the assumption about physical similarity no longer holds. So we cannot completely trust what infants, animals and robots say about their internal conscious states. This makes them unsuitable subjects for identifying the relationships between consciousness and the physical world [Gamez, 2014b].

First-person reports about consciousness from normal adult humans can be combined with measurements of the brain to identify neural correlates of consciousness [Koch *et al.*, 2016]. We can use this data to develop mathematical descriptions of the relationships between consciousness and the physical world. These theories generate descriptions of consciousness from descriptions of physical states, and vice versa, as shown in Fig. 5.

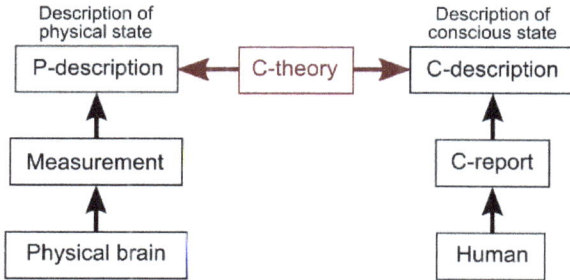

Fig. 5. A mathematical theory of consciousness (c-theory) describes the relationship between physical and conscious states. It can generate a description of consciousness from a description of a physical state and generate a description of a physical state from a description of consciousness.

Tononi's information integration theory (IIT) is an example of a mathematical theory of consciousness [Tononi, 2008]. However, IIT is based on subjective information [Gamez, 2016] and only performs a one way mapping from information to conscious states.

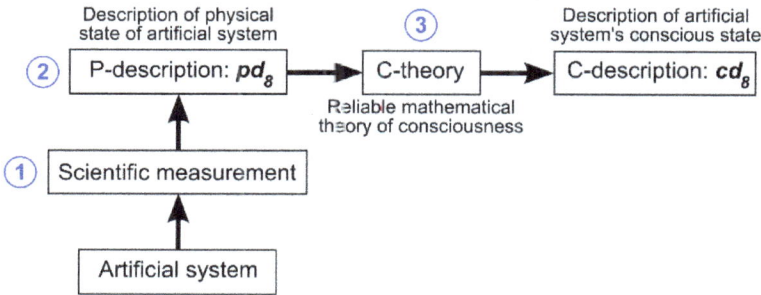

Fig. 6. Deduction of the consciousness of an artificial system. 1) Scientific instruments are used to measure the physical state of the artificial system. 2) Physical measurements are converted into a description of the artificial system's physical state. 3) Reliable mathematical theory of consciousness (c-theory) converts the physical description into a description of the artificial system's consciousness.

When a mathematical theory of consciousness has been judged to be a reliable way of mapping between physical and conscious states, we can use it to make *deductions* about the consciousness of animals and artificial systems. These are deductions, rather than predictions, because they cannot be confirmed by measuring consciousness through first-person reports (see Gamez [2018]). Figure 6 illustrates how a mathematical theory of consciousness can be used to make a deduction about the consciousness of an artificial system.

4.4. *Artificial Consciousness*

A substantial amount of academic research has been carried out on artificial consciousness and working systems have been built to explore different aspects of this topic. Public awareness has been raised through films and TV series, such as *Chappie, Ex Machina, Altered Carbon* and *Black Mirror* [Gamez, 2020]. Artificial consciousness is a complicated field that can be broken down into four overlapping areas [Gamez, 2018]:

- **MC1**. *Machines with the same external behavior as conscious systems.* Humans behave in particular ways when they are conscious. For example, they are alert, they can respond to novel situations, they can inwardly execute sequences of problem-solving steps and they can learn. MC1 machine consciousness is the creation of AI systems that exhibit some or all of these external behaviors. Watson [Ferrucci, 2012] is an example of a MC1 system that mimics the external behavior of conscious humans when they are playing Jeopardy.
- **MC2**. *Models of the correlates of consciousness.* Theories about the neural and functional correlates of consciousness in humans can be modeled in a computer. For example, global workspace implementations have been used to control a naval dispatching system [Franklin, 2003] and a video game avatar [Gamez *et al.*, 2013].
- **MC3**. *Models of consciousness.* Conscious experiences have characteristic features that can be modeled in computers and used to control robots. One example of this type of system was developed by Chella *et al.* [2007], who used a virtual environment (analogous to the robot's consciousness) to control a museum guide robot. Gravato Marques and Holland [2009] built a system in which a robot used a simulation of itself to solve a motor control problem and executed the solution with its real body.

- **MC4**. *Machines that experience conscious states*. When humans are conscious they are immersed in a bubble of experience that contains colors, smells, sounds, etc. (see Sec. 4.1). A machine that was immersed in a bubble of experience, which contained something similar to our colors, smells and sounds, would be MC4 conscious. MC4 consciousness will only be fully solved when we have discovered a mathematical theory of consciousness that can reliably map between physical and conscious states (see Sec. 4.3). We have no idea whether any of our current machines are MC4 conscious.

These categories are not exclusive: systems can implement several of them at the same time. For example, a robot based on the neural correlates of consciousness (MC2) could be phenomenally conscious (MC4) and exhibit conscious external behavior (MC1).

5. Relationships Between Intelligence and Consciousness

5.1. *Natural Intelligence and Natural Consciousness*

Intelligence is a *functional* property: the amount of intelligence in a system is independent of the way in which it is implemented. In Sec. 4.2, I outlined good reasons for thinking that consciousness is linked to spatiotemporal patterns in specific physical materials. Intelligence and consciousness can overlap when the implementation of the intelligence functions produces spatiotemporal physical patterns (for example, neuron firing patterns) that are correlated with consciousness.

While there has been a substantial amount of work on the neuroscience of intelligence [Haier, 2017] and on the neural correlates of consciousness [Koch *et al.*, 2016], we do not know enough about either to be able to say whether the brain's implementation of the functions linked to intelligence are the same as the neural correlates of consciousness. The best we can say is that some of the functions that have been proposed to be linked to consciousness in the brain are also likely to be linked to intelligence. For example, Aleksander and Dunmall [2003] claim that depiction, imagination, attention, planning and emotion are minimally necessary to support consciousness. These functional properties are clearly connected with intelligence — for example, we need imagination to do IQ tasks, such as Ravens' matrices, and planning is related to predictive intelligence and goal achievement. Other people have hypothesized that the brain's implementation of a global workspace is connected with its consciousness [Dehaene, 2014]. Global workspace theory has been shown to be good way to implement AI systems [Franklin, 2003; Gamez *et al.*, 2013], so if global workspace theory is a correct theory of consciousness, then the brain's implementation of a global workspace is likely to be linked to its intelligence. While the exact relationship between prediction and consciousness is an open question, there is clearly a lot of non-conscious prediction going on in the brain, so there is unlikely to be exact alignment between the brain's predictive abilities and its consciousness. More abstract theories about consciousness, such as higher order thought [Rosenthal, 1986], recurrent processing [Maia and Cleeremans, 2005] and information integration theory [Tononi, 2008] point to brain mechanisms that might also be involved in intelligence. For example, a brain that can integrate more information (possibly using recurrent connections) and which contains meta-information about its internal states is likely to be more intelligent. Intelligence can be implemented in many different ways, so there is unlikely to be a strong relationship between the spatiotemporal patterns linked to consciousness and the intelligence functionality of the brain.

Weak inferences can also be made from phenomenological observations about consciousness to the potential intelligence of a system. This connection is weak because most of the data and functions that produce intelligence are not consciously experienced. For example, when an idea spontaneously appears to me, I typically lack insight into the exact mechanisms by which it was arrived at, presumably because it was the result of unconscious processing. However, some of our reasoning is carried out consciously using imagination. With this type of reasoning, a consciousness with more contents could potentially solve more problems, achieve more goals, and generate more predictions. So we might have weak grounds for believing that a system with more conscious contents has greater potential for intelligence. This is only a weak inference because there could be systems with rich states of consciousness that are not capable of intelligent behavior, and an impoverished binary consciousness that only contains 1 or 0 could potentially create every single document that has ever been written by humans. While the intensity of conscious contents plays a role in tagging states as online or offline, this does not appear to be strongly linked to intelligence.

5.2. *Artificial Intelligence and Artificial Consciousness*

The relationships between artificial intelligence and artificial consciousness vary with the type of artificial consciousness.

MC1 machines behave in a similar way to conscious humans. Many external behaviors linked to consciousness are also linked to intelligence, and most of the behaviors that we judge to be intelligent in humans can only be carried out consciously. So there is likely to be a close relationship between progress in MC1 machine consciousness and progress in artificial intelligence. As machines mimic more human behaviors, they will appear to be more conscious and more intelligent. However, there is also a potential dissociation between MC1 machine consciousness and AI. Machines could implement forms of intelligence that achieve low IQ or g-scores on human test batteries, but score highly on universal measures of intelligence. These highly intelligent machines might not exhibit any conscious human behaviors.

MC2 and MC3 machine consciousness research uses models of the correlates of consciousness and models of consciousness to produce more intelligent machines. This has already led to the development of systems that exhibit human-like behavior [Gamez *et al.*, 2013] and intelligent navigation of a museum environment [Chella *et al.*, 2007]. In the future, MC2 and MC3 research is likely to lead to more advanced forms of artificial intelligence. However, AI is a very diverse field and MC2 and MC3 are only two ways of building intelligent machines. A large number of other AI approaches, such as deep neural networks, can be used to develop intelligent systems, and these have few connections to research on consciousness.

We know almost nothing about the MC4 consciousness of artificial systems. It is possible that some of our current AI systems have conscious states that are as rich and vivid as our own. It is also possible that consciousness is only linked to systems that implement certain functions in something approximating biological hardware. Since consciousness is not a purely functional property and a given piece of intelligent behavior can be implemented in an infinite number of different ways [Putnam, 1988], there is not a necessary connection or nomological law linking intelligence and MC4 consciousness. The amount of overlap between MC4 machine consciousness and AI is an *empirical* question that can only be answered when we have a reliable mathematical theory of consciousness and a practical universal measure of intelligence that does not depend on batteries of anthropocentric tests.

6.   Conclusions

Many overlapping definitions of intelligence have been put forward, which are mostly based on human intelligence and generalize poorly to non-human animals and artificial systems. To address this issue, this paper has put forward four hypotheses that define intelligence in terms of prediction.

Progress has been made with the measurement of intelligence in natural systems and many scientists believe that g-score correlates with intelligence in humans and some animals. However, the test battery approach that is used to measure IQ and g-score in natural systems is unlikely to be generalizable to the wide variety of behaviors and intelligences of artificial systems. One solution to this problem is to design tests that only measure human-like intelligence — in the AI context this is known as Turing testing. Another approach is to design universal intelligence measures that can be applied to any system at all, such as the prediction-based measure that was outlined in Sec. 3.4.

Most people agree that consciousness is the stream of colorful, noisy smelly experiences that starts when we wake up in the morning and ceases when we fall unconscious at night. In my own work I have described this as a bubble of experience. Many of the philosophical problems with consciousness can be neutralized with assumptions that provide a reasonable starting point for the scientific study of consciousness [Gamez, 2018]. These assumptions enable us to measure consciousness through first-person reports in normal adult humans. We can then carry out experiments that measure consciousness, measure the physical world, and identify relationships between these two sets of measurements. Scientific research has already made considerable progress identifying some of the neural correlates of consciousness. In the future, we need to discover mathematical descriptions of the relationships between consciousness and the physical world. These can be used to make deductions about the consciousness of non-human animals and artificial systems.

Intelligence is a purely functional property; consciousness is not, so there cannot be a strong connection between consciousness and the many different ways in which intelligence can be implemented in artificial and natural systems. In natural systems, the spatiotemporal physical patterns linked to consciousness might overlap with the brain's implementation of intelligence. Weak inferences can also be made from the richness and structure of natural consciousness to the potential intelligence of a system. In artificial systems there is a reasonably close connection between MC1 machines and machines that exhibit human-like intelligent behavior. MC2 and MC3 technologies can be good ways of building more intelligent machines that think in a similar way to humans.

At the present time we do not have the theories or the data to make stronger conclusions about the relationships between intelligence and consciousness. We will be able to systematically study this relationship when we have a practical universal measure of intelligence, which can be applied to natural and artificial systems, and a reliable mathematical theory of consciousness that can map between physical descriptions and descriptions of conscious states.

References

Aleksander, I. and Dunmall, B. [2003] Axioms and Tests for the Presence of Minimal Consciousness in Agents, *Journal of Consciousness Studies* **10**(4–5), 7–18.

Baker, M. [2016] 1,500 Scientists Lift the Lid on Reproducibility, *Nature* **533**(7604), 452–454.

Bartholomew, D. J. [2004] *Measuring Intelligence: Facts and Fallacies* (Cambridge University Press, Cambridge).

Bell, T. C., Cleary, J. G. and Witten, I. H. [1990] *Text Compression* (Prentice-Hall, Englewood Cliffs).

Bishop, J. M. [2009] A Cognitive Computation Fallacy? Cognition, Computations and Panpsychism, *Cognitive Computation* **1**, 221–233.

Bowers, J. S. and Davis, C. J. [2012] Bayesian Just-So Stories in Psychology and Neuroscience, *Psychological Bulletin* **138**(3), 389–414.

Boysen, S. T. and Capaldi, E. J. [1992] *The Development of Numerical Competence: Animal and Human Models* (L. Erlbaum Associates, Hillsdale, N.J.).

Burkart, J. M., Schubiger, M. N. and van Schaik, C. P. [2017] The Evolution of General Intelligence, *Behav Brain Sci* **40**, e195.

Cattell, R. B. [1971] *Abilities: Their Structure, Growth, and Action* (Houghton Mifflin, Boston).

Chella, A., Liotta, M. and Macaluso, I. [2007] Cicerobot: A Cognitive Robot for Interactive Museum Tours, *Industrial Robot: An International Journal* **34**(6), 503–511.

Chollet, F. [2019] On the Measure of Intelligence. *arXiv preprint* arXiv:1911.01547.

Clark, A. [2016] *Surfing Uncertainty: Prediction, Action, and the Embodied Mind* (Oxford University Press, Oxford).

Cleeremans, A. [2005] Computational Correlates of Consciousness, *Progress in Brain Research* **150**, 81–98.

Collaboration, O. S. [2015] Estimating the Reproducibility of Psychological Science, *Science* **349**(6251), aac4716.

Crosby, M., Beyret, B. and Halina, M. [2019] The Animal-Ai Olympics, *Nature Machine Intelligence* **1**, 257.

Dehaene, S. [2014] *Consciousness and the Brain: Deciphering How the Brain Codes Our Thoughts* (Penguin, New York).

Doya, K., Ishii, S., Pouget, A. and Rao, R. P. N., Eds. [2007]. Bayesian Brain: Probabilistic Approaches to Neural Coding. Cambridge, Mass., MIT.

Ferrucci, D. A. [2012] Introduction to "This Is Watson", *IBM Journal of Research and Development* **56**(3.4), 1:1–1:15.

Franklin, S. [2003] Ida - A Conscious Artifact?, *Journal of Consciousness Studies* **10**(4–5), 47–66.

Gamez, D. [2014a] Can We Prove That There Are Computational Correlates of Consciousness in the Brain?, *Journal of Cognitive Science* **15**(2), 149–186.

Gamez, D. [2014b] The Measurement of Consciousness: A Framework for the Scientific Study of Consciousness, *Frontiers in Psychology* **5**, 714.

Gamez, D. [2016] Are Information or Data Patterns Correlated with Consciousness?, *Topoi* **35**(1), 225–239.

Gamez, D. [2018] *Human and Machine Consciousness* (Open Book Publishers, Cambridge).

Gamez, D. [2019] The Intelligence of Sheep, *Animal Sentience* **25**(27).

Gamez, D. [2020] Consciousness Technology in Black Mirror: Do Cookies Feel Pain?, in *Black Mirror and Philosophy*, edited by D. K. Johnson (Wiley Blackwell, Hoboken), 273–281.

Gamez, D., Fountas, Z. and Fidjeland, A. K. [2013] A Neurally-Controlled Computer Game Avatar with Human-Like Behaviour, *IEEE Transactions on Computational Intelligence and AI In Games* **5**(1), 1–14.

Gardner, H. [2006] *Multiple Intelligences: New Horizons* (Basic Books, New York).

Gravato Marques, H. and Holland, O. [2009] Architectures for Functional Imagination, *Neurocomputing* **72**(4–6), 743–759.

Haier, R. J. [2017] *The Neuroscience of Intelligence* (Cambridge University Press, Cambridge).

Hameroff, S. and Penrose, R. [1996] Orchestrated Reduction of Quantum Coherence in Brain Microtubules: A Model for Consciousness?, *Mathematics and Computers in Simulation* **40**, 453–480.

Harnad, S. [1994] Levels of Functional Equivalence in Reverse Bioengineering: The Darwinian Turing Test for Artificial Life, *Artificial Life* **1**(3), 293–301.

Hernández-Orallo, J. [2000] Beyond the Turing Test, *Jounal of Logic, Language and Information* **9**(4), 447–466.

Hernández-Orallo, J. [2017] Evaluatior in Artificial Intelligence: From Task-Oriented to Ability-Oriented Measuremen, *Artificial Intelligence Review* **48**(3), 397–447.

Hernández-Orallo, J. and Dowe, D. L. [2010] Measuring Universal Intelligence: Towards an Anytime Intelligence Test, *Artificial Intelligence* **174**, 1508–1539.

Hingston, P. [2009] A Turing Test for Computer Game Bots, *IEEE Transactions on Computational Intelligence and AI In Games* **1**(3), 169–186.

Humphreys, L. G. [1979] The Construct of General Intelligence, *Intelligence* **3**, 105–120.

Husserl, E. [1964] *The Phenomenology of Internal Time-Consciousness* Translated by J. S. Churchill. (Martinus Nijhoff, The Hague).

Hutter, M. [2021]. The Hutter Prize. Retrieved 12/11/21, 2021, from http://prize.hutter1.net/.

Knill, D. C. and Pouget, A. [2004] The Bayesian Brain: The Role of Uncertainty in Neural Coding and Computation, *Trends Neurosci* **27**(12), 712–719.

Koch, C., Massimini, M., Boly, M. and Tononi, G. [2016] Neural Correlates of Consciousness: Progress and Problems, *Nat. Rev. Neurosci.* **17**(5), 307–321.

Legg, S. and Hutter, M. [2007a]. A Collection of Definitions of Intelligence. *Proceedings of Advances in Artificial General Intelligence Concepts, Architectures and Algorithms: Proceedings of the AGI Workshop 2005*, edited by B. Goertzel and P. Wang, IOS Press, pp. 17–24.

Legg, S. and Hutter, M. [2007b] Tests of Machine Intelligence, in *50 Years of Artificial Intelligence*, edited by M. Lungarella, F. Iida, J. Bongard and R. Pfeifer (Springer, Berlin, Heidelberg), 232–242.

Legg, S. and Hutter, M. [2007c] Universal Intelligence: A Definition of Machine Intelligence, *Minds and Machines* **17**, 391–444.

Maia, T. V. and Cleeremans, A. [2005] Consciousness: Converging Insights from Connectionist Modeling and Neuroscience, *Trends in Cognitive Sciences* **9**(8), 397–404.

Marino, L. and Merskin, D. [2019] Intelligence, Complexity, and Individuality in Sheep, *Animal Sentience* **25**(1), 1–26.

Neff, G. and Nagy, P. [2016] Talking to Bots: Symbiotic Agency and the Case of Tay, *International Journal of Communication* **10**, 4915–4931.

Neisser, U., Boodoo, G., T. J. Bouchard, J., Boykin, A. W., Brody, N., Ceci, S. J., Halpern, D. F., Loehlin, J. C., Perloff, R., Sternberg, R. J. and Urbina., S. [1996] Intelligence: Knowns and Unknowns, *American Psychologist* **51**(2), 77–101.

Pockett, S. [2000] *The Nature of Consciousness: A Hypothesis* (Writers Club Press, San Jose, California).

Portugues, R., Severi, K. E., Wyart, C. and Ahrens, M. B. [2013] Optogenetics in a Transparent Animal: Circuit Function in the Larval Zebrafish, *Curr Opin Neurobiol* **23**(1), 119–126.

Putnam, H. [1988] *Representation and Reality* (MIT Press, Cambridge, Massachusetts; London).

Robertson, K. F., Smeets, S., Lubinski, D. and Benbow, C. P. [2010] Beyond the Threshold Hypothesis: Even among the Gifted and Top Math/Science Graduate Students, Cognitive Abilities, Vocational Interests, and Lifestyle Preferences Matter for Career Choice, Performance, and Persistence, *Current Directions in Psychological Science* **19**(6), 346–351.

Rosenthal, D. M. [1986] Two Concepts of Consciousness, *Philosophical Studies* **49**(3), 329–359.

Sanghi, P. and Dowe, D. L. [2003] A Computer Program Capable of Passing Iq Tests. 4th Intl. Conf. on Cognitive Science (ICCS'03). Sydney: 570–575.

Shaw, R. C. and Schmelz, M. [2017] Cognitive Test Batteries in Animal Cognition Research: Evaluating the Past, Present and Future of Comparative Psychometrics, *Animal Cognition* **20**, 1003–1018.

Sternberg, R. J. [1985] *Beyond Iq: A Triarchic Theory of Human Intelligence* (Cambridge University Press, Cambridge).

Thurstone, L. L. [1938] *Primary Mental Abilities* (University of Chicago Press, Chicago).

Tononi, G. [2008] Consciousness as Integrated Information: A Provisional Manifesto, *Biological Bulletin* **215**(3), 216–242.

Turing, A. [1950] Computing Machinery and Intelligence, *Mind* **59**, 433–460.

Warwick, K. [2000] *Qi: The Quest for Intelligence* (Piatkus, London).

Chapter 7

Attention and Consciousness in Intentional Action: Steps Toward Rich Artificial Agency

Paul Bello*, Will Bridewell

Code 5512, Naval Research Laboratory, 4555 Overlook Ave. S.W., Washington, District of Columbia 20375, USA
**paul.bello@nrl.navy.mil, will.bridewell@nrl.navy.mil*

If artificial agents are to be created such that they occupy space in our social and cultural milieu, then we should expect them to be targets of folk psychological explanation. That is to say that their behavior ought to be both predictable and explicable in terms of beliefs, desires, obligations, and especially intentions. Herein we focus on the concept of intentional action, and especially its relationship to consciousness. After outlining some lessons learned from philosophy and psychology that give insight into the structure of intentional action, we find that attention plays a critical role in agency, and indeed in the production of intentional action. We argue that the insights offered by the literature on agency and intentional action motivate a particular kind of attention-centric computational cognitive architecture that we have used as a blueprint for the design of the ARCADIA cognitive system. We detail the inner workings of ARCADIA and demonstrate their relevance in capturing the tricky relationship between attention, intention, and consciousness via a computational exploration of cases of so-called causally-deviant action.

Keywords: attention; intention; consciousness; agency

1. Introduction

The term "agent" is a central fixture in the literature on artificial intelligence (AI), no matter if written for general or academic audiences. AI researchers design their software to engage in activities such as labeling images and driving cars, and agency is fundamentally concerned with action. Perhaps the most influential definition of an AI agent is taken from Russell and Norvig (2010), who defines an agent as, "Anything that can be viewed as perceiving its environment through sensors and acting upon that environment through actuators." Those authors further claim that the field of AI research essentially concerns the study and design of rational agents, which are exclusively those agents who "act so as to maximize the expected value of a performance measure based on past experience and knowledge." Various other definitions invoke learning and flexible adaptation as a key feature of intelligent agents, and still others demand that agents have the representational repertoire to reason about beliefs, desires, and intentions (Padgham and Winikoff, 2005). All these definitions point to a specific feature of action: its relation to reasons. Whether maximizing a performance measure or forming intentions, an agent does both of these on the basis of beliefs and desires, which may be broadly construed either logically or decision-theoretically. Absent any further qualifications, however, such simple guiding principles for agency lead to pitfalls, including the development of plans that instrumentally maximize utility while doing harm, unanticipated by the designers, in the process of achieving goals. Furthermore, advances in machine learning have led to systems that demonstrate impressive successes on narrow classes of problems. These systems may generate a strong impression of agency even though they are often based on relatively inexpressive representational formalisms and are unable to provide human-understandable explanations for their successes and failures.

In the past, we have expressed pessimism about the sufficiency of the *appearance of agency* in grounding *ascriptions of agency* (Bello and Bridewell, 2017). On our view, at least three different classes of agents can be identified, with contemporary efforts in AI and computational cognitive science having addressed only two of these. The first class of agents is exemplified by reinforcement learning agents[1] that are trained to behave with respect to known features of the world, possibly including known descriptions of uncertainty. In this class, agent behavior is determined by optimizing the choice of action in every state of the world, which yields fixed

---

[1]We only intend to use a simple conception of RL agents here to make our point. Of course RL agents can and have been augmented to perform tasks that are more complex. However, many of these additions, such as adding a memory to an agent and taking it to be a part of the environmental state, are only tractable for extremely limited environments and temporal horizons. Additionally, since RL agents almost entirely are defined using a knowledge representation language that is less expressive than first-order logic it remains to be seen how information stored in these memories might be used for sophisticated reasoning about causes, or the mental states of other agents, which require forms of knowledge representation beyond first-order logic to capture with fidelity. On the other hand, RL as a technique has been used very successfully as part of the second class of agents we discuss.

policies that function well in a world of knowns. If these kinds of agents encounter conflicts among applicable actions, these are resolved by optimizing for expected value over time. The second class of agents is harder to describe but uniquely involves a generative mechanism for considering past, future, and counterfactual states of the world when selecting among actions. These agents clearly violate the kind of Markovian assumptions present in the first class of agency via reliance on memory, and in the case of counterfactuals, on information that may not be described by any prior distribution in the agent's possession. Conflicts in these sorts of agents are often resolved by way of explicitly represented preferences, or through using their generative mechanism for problem-solving when the conflict results from an impasse due to the agent's current state representation.

Both types of agents discussed so far share a glaring deficit with respect to intelligent agency: they accomplish their purposes without properly *doing* anything at all. Although this may sound like an outlandish or overblown claim, the point of this chapter is to make the case for agent-doings as paradigmatic cases of intentional action. These sorts of actions are to be considered in contradistinction to reflexes, habits, and automatisms that are detached from occurrent intentions. The view defended here is that intentions contribute substantially to the control of action through attention (Bello and Bridewell, 2017; Wu, 2016; Watzl, 2017) and that agents contribute to the adoption and maintenance of their intentions over time through practical reasoning. This marriage of intentions and attention is novel with respect to prior work on modeling intentions in artificial systems (Cohen and Levesque, 1990; Rao and Georgeff, 1991) and, as will be argued, is crucial for bridging the gap between intentions and intentional action. This configuration leaves room for the agent to act, or at minimum to guide action, and thus to be a target for responsibility for subsequent outcomes.

The motivation for understanding intentional action is clear: many of our most important socio-cognitive practices including blame and praise depend on ascertaining the intentional status of actions. Fleshing out the details of an attention-centric notion of intentions raises questions about the role of consciousness in agency When action outcomes begin to deviate from intentions in everyday life, it may often lead to revisions in judgment about the actor's degree of agency, control, or skill in performing that task. Each of these can manifest as a conscious appraisal and serve as top-down modulation for currently active intentions. Progress on intentions may in turn lead to progress toward machine consciousness, albeit in a very limited sense. The plan for the rest of the chapter will proceed as follows. Two vignettes will be explored that will help distinguish between intentionally acting and merely behaving. A brief analysis of these vignettes will lead to a set of constraints on the design of a computational cognitive architecture that has the capacity for agency. After compiling a list of architectural constraints and some desiderata for how intentions might be represented and deployed, we present the ARCADIA cognitive system, which has been designed with respect to many of these constraints. An implementation of one of the vignettes will be provided using ARCADIA. We conclude with a discussion of the areas where cognitive fidelity may be improved in both the model and the cognitive system if they are to resemble human-like agency.

2. Agency and Intentional Action

Ultimately, this chapter concerns itself with the notions of agency and intentional action, and of course their relation to one another. Rather than providing definitions up front, we present two scenarios to tease out important distinctions between various forms of behavior with respect to agency. The second of the pair will review a well-worn thought experiment in the philosophy of action, whereas the first is grounded in real-world events.

2.1. *Kenneth Parks*

One evening, a somnambulant Kenneth Parks arose from the couch, dressed himself, got in his car, drove 14 miles, and navigated three different traffic lights to arrive at his in-laws' home. Once there, he walked into the house, strangled his father-in-law into unconsciousness, and stabbed his mother-in-law multiple times in the chest and shoulder. She later died of her wounds. Mr. Parks left his in-laws' house and drove to the police station, where he told the authorities that he thought that he had hurt someone. He was horrified to find out what he had done. His memories were fragmentary, with only images of watching TV on his couch, his mother-in-law's horrified face, the knife in his hand, and a few other highly emotional bits and pieces, but nothing connecting the pieces together. Parks was charged with first-degree murder and pled not guilty relying on the defense of "non-insane automatism," which in layman's terms is effectively sleepwalking. Parks had a history of disturbed sleep, both walking and talking in his sleep, as well as sitting up in bed with his eyes wide open yet being totally unresponsive. Parks was later assessed as displaying parasomniac tendencies, and the various stresses he had been under served as a good explanation of why this particular episode occurred. The fragmentary structure of his recollections of the incident was also consistent with sleepwalking as an explanation. After an extensive investigation, Parks was acquitted of murder, and the acquittal was upheld by the Supreme Court of Canada (Broughton *et al.*, 1994). The lynchpin to the judgment lies in the dependency between "being aware of what you are doing" and acting intentionally.

Kenneth Parks' case provides a vivid study of how complex behavior is possible in the absence or near-absence of conscious awareness. If he had woken up while driving to his in-laws, he would have presumably had access to background knowledge, norms, and autobiographical episodic memories of his in-laws, which by his own post hoc accounting were positive. This information, along with the capacity for conscious control of action, likely would have saved their lives. Based on this counterfactual description, there are three general, salient characteristics that stand out: the disconnect between consciously mediated behavior and script-like automatisms, the notion that specifically conscious information widely recruits background knowledge and memories in service of deliberation, and the idea that attention enables exercises of control. These properties are a good fit with a popular theory of consciousness: the Global Workspace Theory (GWT) (Baars, 1997; Dehaene *et al.*, 2006). Proponents of GWT see the mind as consisting of many distributed modules that actively shape the contents of short-term stores that are broadly available for use in reasoning, decision-making, and other high-level cognitive activities. The contents of this workspace are cyclically "broadcast" throughout the mind, and coalitions among the modules are formed to promote their processing output into the workspace in the immediate future. Kenneth Parks' case was explored and explained in some detail from the perspective of GWT by Neil Levy (2014) in his book on consciousness and moral responsibility. On his analysis, if Mr. Parks had been sufficiently consciously aware, his conscious contents would have been integrated via broadcast with his dispositions and other memories. Since Levy's book is about responsibility, the conclusion he draws is that Mr. Parks is not responsible for his actions because he was unable to recruit the background knowledge about their moral status. Consequently, he had no way to grasp their implications and thus to be responsive enough to reasons such that he might have done otherwise. On this account, the roles of attention and subsequently consciousness are as facilitators for practical reasoning, which yields the intentions that accompany normal action. Is this the extent of attention's involvement with intentional action, or is there more to the story?

2.2. *Deviant Causal Chains*

Consider next a fictional vignette based on an example from Chisholm (1964). After an argument with his nephew Smith, Jones retires to his bedroom and slips under the covers with Mary, his new girlfriend. Earlier that day, Smith, who has been taking care of Jones in his old age, cautioned Jones about all the expensive items that Mary has been purchasing with his money. Looking at Mary, Jones becomes angry with Smith for thinking she is taking advantage of his wealth. At that moment, he decides to give Smith's inheritance to Mary. A week later, Smith receives a letter announcing that he has been excised from Jones' will and that his services would no longer be required. Smith flies into a rage and decides to drive to his uncle's house and kill him. Smith grabs his shotgun, gets into his car, and begins to drive. On his way, Smith mentally replays his last conversation about Mary, and thinks about his loss of an inheritance after all the time and energy he had spent on caring for old Jones. Soon, Smith finds himself in a feverish rage, his thoughts preoccupied with the bloody work he is about to do. Then, all of the sudden, the car slams into something. Smith is broken from his dark fantasies to find that he has run over a pedestrian. He exits the car to investigate and, to his surprise, finds that he has run down his uncle Jones.

The story of Smith killing Jones provides a window into another aspect of the consciousness–agency nexus. The standard way of thinking about agency in philosophy is in terms of the causal theory of action. The causal theory can be understood as claiming that actions are events caused by an agent's mental states (e.g., beliefs, desires, and intentions). This view is not only standard among philosophers but also a foundational one in AI, yet it has a glaring problem: agentive control. It is assumed that the causal relationship between intentions and actions are exercises of agentive control, yet causal deviance examples such as with Smith and Jones militate against the view that intentions directly cause action. Presumably, if Smith had been firmly focused on driving, he would have noticed the cars, street signs, and pedestrians around him. So it would seem that Smiths' failure to exert control in order to keep his intention to drive to Jones' house firmly in the front of his mind led to his unawareness of the pedestrian and eventually to an unintentional killing. Wu (2016) has suggested that in not accounting for the role of attention in intentional action, the causal theory is unable to deliver an adequately explanatory concept of agentive control. He and others defend a theory of attention called *selection for action*, which broadly put, states that if a subject S selects X to inform the performance of task T, then S attends to X. On this view, attention mediates the connection between having a task or intention in mind and generating intention-consistent outcomes.

2.3. *Taking Stock*

These two vignettes point toward a rather complex view of agency as something not necessarily reducible to differences in knowledge representation from the standard sort of agents to be found in AI or computational cognitive science. However, the differences in architecture and general processes implied by vignettes also point to differences in how key concepts such as intentions ought to be treated if moving toward human-like agency is the objective. The Kenneth Parks case illuminated the complexity of behavior that is possible even in the absence of consciousness. Levy's plausible account of abnormal practical and moral reasoning in the absence of the integrative function of consciousness, mediated by attentional selection, points toward an agent implementation with distributed components that become synchronized during the conscious performance of tasks. The causal deviance story pointed to the close connection between having attentional priorities play a role in how intentions are represented and used to guide behavior. Distributed architecture, synchronization, and attentional priorities are featured centrally in the ARCADIA cognitive system, making it a promising framework for computational research on developing agents with the elusive sort of human-like agency mentioned earlier.

3. The ARCADIA Cognitive System

The previous section mentioned a cognitive architecture consisting of independently operating modules that can be synchronized by the broadcasting of conscious contents and a mechanism for selective attention that interacts with intentions to produce agent-guided intentional action. The ARCADIA cognitive system was designed with consideration given to the primacy of attention in coordinating perception, cognition, and action in a way that supports the development of artificial agents in the sense discussed in the introduction.[2] Given its start as a model of attention (Bridewell and Bello,

---

[2]It should be noted that there are a number of other computational conceptions of attention on offer in the literature, some of which may be able to be repurposed to do the work that attention is doing in ARCADIA in accounting for intentional action. ARCADIA's task-specific attentional priorities are unique in the sense of providing

2016a) ARCADIA has been used to investigate and model human behavior on a number of tasks that constrain theories of attention such as displaying inattentional blindness (Bridewell and Bello, 2016b) and change blindness (Bridewell and Bello, 2015) under conditions similar to those where humans display the same limitations. ARCADIA has also been used to capture human performance data on a variety of psychological lab tasks (e.g., Lovett *et al.*, 2019; Bello *et al.*, 2018; O'Neill *et al.*, 2018; Briggs *et al.*, 2017).

3.1. *Representation: Components and Interlingua*

ARCADIA is technically an architecture schema: in other words, it is a framework within which intelligent systems can be developed. Systems developed with this schema consist in a non-empty set of components that carry out information processing, a focus of attention that guides the computation within components, and a routine for selecting the focus of attention. Every component must be able to read from and write to a common data format called *interlingua*, an example of which looks like the following:

```
{:name "visual-object"
:arguments {:color "Red" :shape "Rectangle"
            :image matrix-image<34x56>}
:world "working-memory"
:source working-memory
:type "instance"}
```

An interlingua element has a name, which identifies a general relation among properties; an argument, which are property–value pairs; a world, which enables the separation of real, hypothetical, historical, and other information; a source, which records the element's provenance; and a type, which can be used to express a type/token distinction (as shown here) or to be used as a logical *sort*.

ARCADIA is a representationally heterogeneous system: the internal processing of components can be implemented by anything that computes a function from the component inputs to the desired output, potentially including symbolic, neural, and probabilistic data structures and algorithms. Cognitive modelers might choose to implement components and their

---

top-down guidance of attention that remains open to within-task learning, potential for changes in objectives, environmental dynamism, and a direct connection to inhibition and other forms of attentional control mechanisms. Much of this functionality is unreported in this chapter.

internal processes in a manner that is informed by literature in neuroscience and psychology whereas an AI researcher might be under no other constraints but to build the most efficient algorithm possible for the task by using few components and cognitively implausible internal processes. The most complex ARCADIA models that have been developed have relied on a mixture of cognitively plausible components and less-constrained components where there is a dearth of empirical data to inform their design. Fundamentally, if components are seen as simple functions that can take previous outputs of other components and process them, then ARCADIA can be seen as a simple model of computation that uses focus-selection routines to compose functions over cycle time.

3.2. *The Cycle*

During each ARCADIA cycle, components will perform internal processing, framing their results as interlingua elements as shown above. These results are made available for use to all components in the subsequent cycle. The interlingua is designed to enable component interoperability, and because it supports heterogeneous representations in its arguments field, multiple components can process information from different parts of the same element. For instance, a component implementing a neural network can classify image data, while a reasoning component can process the symbolic elements of the same argument list. On each cycle, ARCADIA selects a new focus of attention from the set of available component outputs, and gives that element primacy among all the others in the subsequent cycle. This means that components can be designed to respond to the focus when they can and to engage in default behavior when they cannot. We refer to components that give priority to the focus of attention when possible as *focus responsive* and those that engage in the same sort of processing regardless of the focus as *focus unresponsive*. Focus-unresponsive components are good candidates for producing the kind of automatic behaviors that might arise in the absence of having a conscious intention, and thus provide a means of exploring the role of control which was so central to the two vignettes presented earlier.

3.3. *Focus Selection*

On every cycle, ARCADIA is faced with a selection problem because components generate their outputs in parallel. Some components process the input stream at a low level, producing features that need to be integrated into representations of objects, while others might produce overly learned action responses that are driven by the current state of the environment (i.e., the contents of short-term sensory memory components), and others still might produce planned action in accordance with a maintained intention. At various times over the course of processing, ARCADIA may need to select differentially among these component-produced interlingua elements. To influence this selection problem, we specify attentional priorities. As a starting point, ARCADIA maintains a default set of attentional priorities that biases the system toward focusing on potential actions to execute, followed by mid-level representations in the environment (e.g., people and objects), and then by visually salient regions in the input. Notably, the action requests include not only overt physical activity but also memory-related actions, task-switching actions, and other mental operations. What is captured by this admittedly simple representation of attentional biases is the fact that bottom-up pulls on attention from low-level salience computations are washed out whenever task-relevant bias or task-related actions are available to be selected.

3.4. *Tasks and Task-Switching*

Every ARCADIA model can rely on the default attentional strategy, but most of these are models of tasks, and thus have internal representations of task structure. Task representations in ARCADIA are distributed across the system and consist of semantic information about the task, such as its name, along with procedural information in the form of stimulus-response (SR) links that match against the collection of contents produced on each cycle by the model's components and generate action requests (Bridewell *et al.*, 2018).

Along with this standard set of task-related information, each task representation has its own set of attentional priorities that attunes the model to task-relevant features of the environment. When a request to switch tasks is attended to and processed, the task representation is loaded into a section of working memory dedicated to tasks, and the attentional priorities that are part of the task representation are combined with ARCADIA's default strategy to begin biasing attentional selection toward task-relevant features of the environment. The task and associated priorities remain active until a new task is adopted, at which time the priorities of the new task are combined with the default strategy to guide attention.

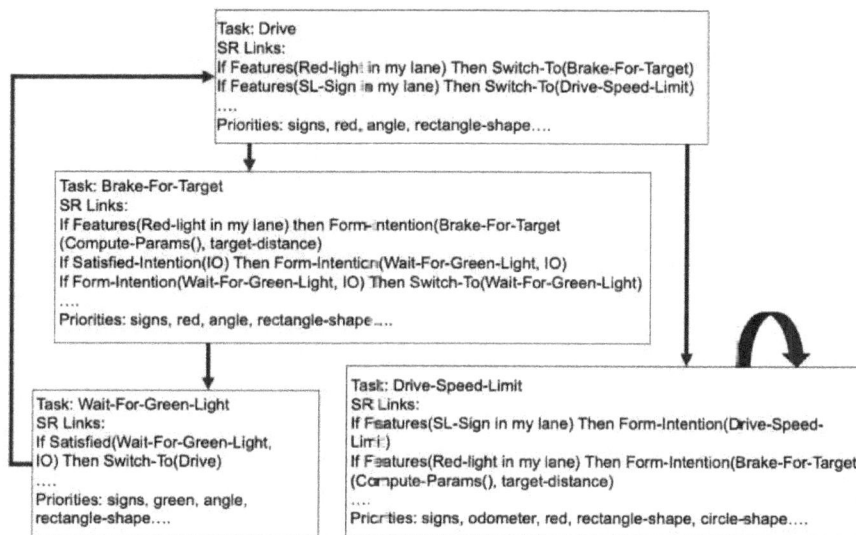

Fig. 1. A high-level representation of some interrelated driving tasks.

3.5. *Intentions*

The internal structure of an intention is given in the snippet of ARCADIA interlingua below. For brevity, the values of the properties in the argument map have been omitted.

```
{:name "activated-intention"
 :arguments
    {:constraints :evaluation-history
     :formed-at :intentional-object
     :precondition-met :satisfied?
     :subtype :task :terminated?}
 :world "intention-memory"
 :source component
 :type "instance"}
```

Following Fig. 2, the precondition-met argument is a boolean value that tracks whether the activation conditions for the intention have been obtained, while similar fields corresponding to the satisfaction and termination conditions are also present. On every cycle, monitoring functions evaluate the status of the available interlingua elements with respect to the constraints that define activation, satisfaction, and termination conditions. As in Fig. 2, changes in these values alter the status of an intention, moving it from being merely maintained to being active and then either satisfied or terminated prior to satisfaction. The subtype field marks whether the intention is an achievement or maintenance intention. Achievement intentions terminate immediately when their satisfaction conditions are met. This corresponds roughly to a successful intentional action. Maintenance intentions, such as driving the speed limit, do not terminate upon satisfaction, rather, they persist until they are supplanted by another maintenance intention that directly conflicts with them, such as driving a different speed limit. The intentional-object argument holds a representation of the particular task-relevant objects and relations toward which the intentional action is directed. As abstract intentions become more specific and directed at particulars, the intentional object field is updated accordingly. Finally, the task field is a semantic marker that links the intention to a task representation and all the corresponding procedural knowledge and attentional priorities contained therein.

As can be seen in Fig. 1, the formation of intentions is a type of mental action, appearing on the right-hand side of various SR links. Action-monitoring information that tells of the satisfaction of a currently active intention can serve as a signal to form new intentions which, environmental conditions depending, may result in the intention's activation and adoption of the corresponding task. Intentions have the interesting property of not necessarily influencing behavior as soon as they are formed. In fact, many of our intentions are future-directed and can be seen as a form of prospective memory, contributing to our capacity to flexibly plan and adapt. Intentions are often specified at an abstract level, such as "intending to go out for lunch," without any other information except for knowing that it will require driving and all the procedural and semantic and navigational knowledge that comes along with that activity. However, being committed to going out to lunch settles the question of whether one stays home. Moreover, given that having an intention commits an agent to whatever goal is specified by the intention, other activities during that time are precluded. Once in the car, habits, memories, perceptions (e.g., of restaurants or billboard advertisements) conspire to give shape to the original abstract intention until a choice has been made and a new intention is formed to go to a specific restaurant. All this occurs in parallel with driving which has its own corresponding intentions associated with route retrieval, planning, and observing traffic regulations. Finally, intentions must sometimes

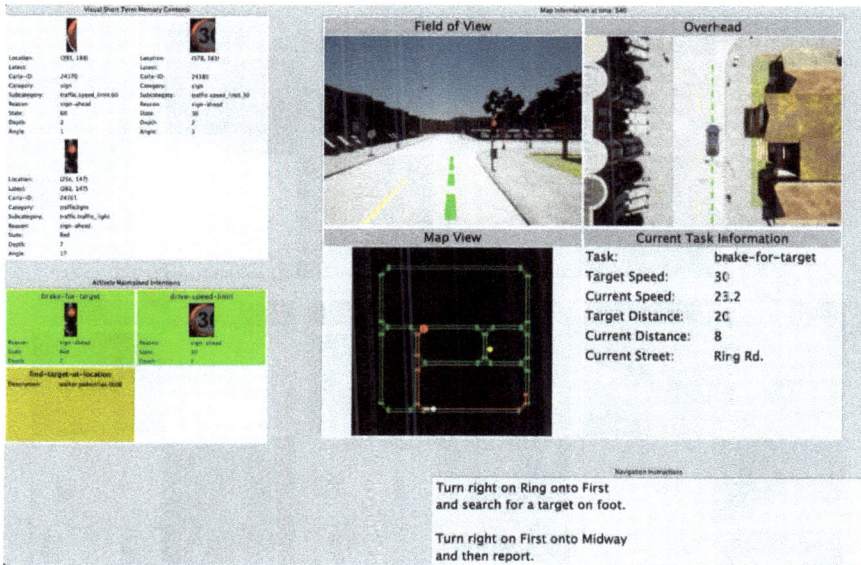

Fig. 2. Intentions (lower left) are formed both via task instructions (lower right) and via procedural knowledge about driving. In the screenshot above, an achievement intention for stopping at a currently red traffic light and maintenance intentions for following the 30 km/h speed limit are currently active, while an achievement intention for finding a target pedestrian remains inactive until the vehicle is on First St.

be maintained to not be forgotten or washed out in a sea of distractions. Along with the practical reasoning involved in choosing which actions become intentions, maintenance leaves room for the agent to do something: to exercise control by intervening in its intentions. This rather complex picture of intentions has been taken as aspirational in the development of intentional action in ARCADIA, however much work remains to achieve it.

4. Attention, Intention, and Causal Deviance: A Computational Model

To illustrate the difference between intentional and unintentional action, an implementation of the Smith/Jones case in ARCADIA was built and further modified to superficially illustrate how intentional action might work in the system. These models comprise close to forty interacting components, including functionality for visual processing, and short and mid-term memory stores.

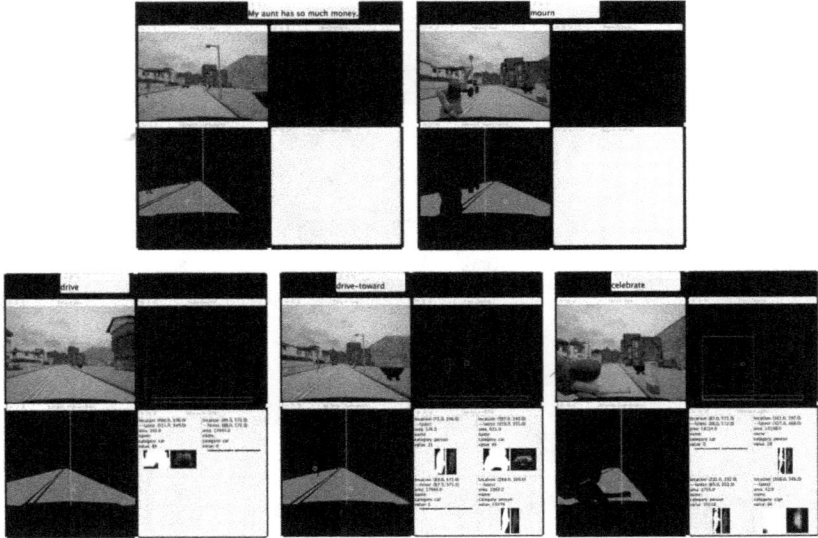

Fig. 3.   Top: Unintentional killing. The model relies on a single point in the periphery to control driving. Attention is focused on the contents of inner speech, and so visual memory (lower right corner of each video) is empty. Bottom: Intentional killing. The model is focused on driving and attentional priorities include cars, signs, and pedestrians as evidenced by the contents of visual memory. Target is identified and marked with a second control point for steering, and a trajectory is planned to run the target down.

As mentioned earlier, some of ARCADIA's components are not focus-responsive. In the models shown in Fig. 3, two simple proportional-integral-derivative (PID) controllers from the throttle and steering wheel produce outputs on every cycle subject to the position of one or more control points (the red dots shown in the lower-left windows). The bottom-most control point is automatically computed and is available to the system without having to attend, however steering based only on this one control point results in erratic driving that fails to take upcoming information into account in the steering control computations. A second control point can be placed on the road ahead or on any other person or object in the visual field to further inform steering, or in the case of the intentional killing sequence in the bottom row, to mark a target to aim the vehicle towards. We stress that the goal of this particular model was to demonstrate salient high-level differences between "intentional" and "unintentional" action. The corresponding gaps in both theoretical and model-related contact reflect the rather limited sense in which these high-level differences were framed during the development of the model.

The model relies primarily on three distinct task representations having their own sets of attentional priorities and procedural knowledge in the form of SR links. The first of these is the "drive" task which, when selected, biases attention toward lane lines, other cars, traffic control devices, and pedestrians. The second task is "rumination" in which attention is biased toward generating inner speech associated with ruminating on a plan. This is accomplished by prioritizing action requests for subvocalizations over other requests. Finally, the "run-over-uncle" task relies on adding extra identifying information to attentional priorities that privilege pedestrians that have the appearance of the uncle.

In the unintentional case highlighted in the top two images of Fig. 3, the model begins executing with "rumination" as its currently selected task. At this point, ARCADIA is focused solely on the contents of inner speech and relies solely on attention-free, one-control-point driving behavior produced by a focus-unresponsive component. The driving is erratic, and the car runs down the pedestrian by accident. Because the attentional strategy associated with rumination prioritizes focus on interlingua elements containing inner speech at the expense of driving and its corresponding attentional priorities, no driving-relevant information from the outside world appears in visual memory (the lower-right corners of each window), meaning that ARCADIA never saw the pedestrian at all. In contrast, the images on the bottom row of Fig. 3 illustrate the "intentional" action of running the uncle over. In this case, the model begins similarly in an erratic pattern of driving and quickly drifts into the left lane. A monitoring component generates an alert that grabs attention and facilitates a task switch back to driving, after which the model begins to attend to the lane, cars, and pedestrians. Once the pedestrian is attended to and recognized as the uncle, a plan is made to run the uncle over. The second control point for driving is fixed on the uncle and used to inform steering the car into him.

5. Limitations and Open Questions

The computational model presented above illustrates a handful of the features of intentions discussed previously without having been implemented using intentions as they currently exist in ARCADIA.

5.1. *Monitoring*

One notable feature of the model presented above was the inclusion of a dedicated component for detecting when the vehicle crossed the midline in the road. When this state of affairs was obtained, the component generated a signal that subsequently grabbed attention, and initiated a switching of tasks back to driving. Even if this seems like a sensible enough strategy, there will be times when the car crosses the midline intentionally in order to achieve some end such as avoiding a car, an animal, a pedestrian, or an object that has fallen into the road. There may be other times when careless driving is tolerable due to a low probability of risk. Admittedly, many details remain to be worked out, but this suggests that monitoring is deeply intertwined with intentions rather than existing separately.

The processes that underwrite monitoring are not computationally well understood. Recall that intentions themselves exist at multiple levels of abstraction, starting with highly abstract specifications such as "drive to lunch" down a level to implementation intentions such as "drive to the Chinese restaurant at the corner of 1st avenue and Park street," and finally to motor intentions to move the wheel and push the pedals. One might imagine that monitoring for different classes of intentions is implemented in qualitatively different ways. Monitoring is just one place where consciousness directly connects to intentional action. Whether human beings are conscious of their motor intentions is a controversial topic but the conservative assumption is that they are not. Insofar as there is an awareness on the agent's part of one of its motor programs failing, it is specified one level up in the intention hierarchy as progress toward satisfying intentions at that level becomes stymied. A similar interaction occurs between the middle and top layers of the hierarchy. Recall that the middle layer contains intentions that are concrete in terms of referencing particular objects, locations, and so on, whereas intentions at the top-most layer are often abstract, specifying only high-level goals and procedural knowledge that eventually gets refined one layer down the hierarchy. Motor programs are plausibly monitored outside of conscious awareness, but the role of consciousness is less clear in the middle and top layers. Exhausting immediately available intentions-to-X in the middle layer of the hierarchy may result in a conscious judgment that one's high-level intentions require reconsideration. This judgment may play a role either in revising high-level goals or in prompting problem-solving or other forms of deliberation.

5.2. *Automaticity and Complex Tasks*

Perhaps the most implausible features of the computational model presented earlier were how routinized driving behavior was accounted for as well as how "control" was regained through no more than an externally cued task-switch. It is widely agreed upon that driving skills become proceduralized and thus executable with drastically less involvement of conscious attention. On well-practiced routes, subjects have been shown to exhibit inattentional blindness that tracked their judgments of the ease of the navigation task. With practice, the task becomes less demanding of attention but also comes at the cost of failing to notice relevant changes in the environment. Empirical data on divided attention in driving yields similar conclusions. But should it be concluded then that subjects lack any and all awareness of the primary task while their attention is divided, or while they are in the autopilot state of mind typical of drivers in low-stress, well-known environments? Unfortunately, there is no clear answer, but this is primarily because most complex tasks such as driving are decomposable into many smaller tasks, some of which require attention and some that ostensibly do not. In driving, for example, studies have been performed on a decomposed list of driving tasks including mirror-checking, gear-changing, following at a distance, approaching intersections, and braking for traffic control devices. In the presence of a secondary task, some of these activities exhibited degraded performance, while others remained stable (Duncan *et al.*, 1993). Even though this says little about the phenomenology experienced by the subject, it does give a sense of which tasks are automatized.

5.3. *Control and the Value of Control*

It is almost certainly true that most complex tasks are also heterogeneous in terms of the amount of controlled processing that they require. As uncertainty, risk, or performance evaluation criteria change a shift toward controlled processing from automaticity becomes more likely. The Smith/Jones thought experiment is unrealistic in the sense that oncoming cars, loud sounds, and the motion of pedestrians on the sides of the road would frequently pull attention away from Smith's dark fantasizing and direct it outward toward the environment frequently enough for him to have something of an ongoing general awareness. Whether Smith attends frequently enough to driving in order to spot Jones in time to avoid a collision is an interesting question that a more realistic model could be built to answer, but building it requires further elaboration of ARCADIA's facilities for representing intentions.

In the abstract, ARCADIA represents control knowledge in the form of attentional priorities, either as built-in default values or as task-specific values associated with whichever task happens to be active. Individual sets of attentional priorities provide top-down guidance for attention while performing a particular task, biasing ARCADIA away from attending to task-irrelevant noise. However, complex tasks involve many subtasks in various stages of completion that must be managed by the agent without degrading overall performance. In Fig. 2 we see a braking task and a speed-limit maintenance task that are simultaneously active, one of which is trying to slow the vehicle to a stop while the other is trying to maintain a non-zero speed. At the present time, there is no principled way in ARCADIA to detect and manage such conflicts effectively. Both intentions produce requests to act, but the intention associated with the currently adopted task wins out since its attentional priorities will bias attentional selection toward actions specific to the current task. This scheme has only limited utility, especially for cases that lack conflict among multiple active intentions and where multitasking behavior is desirable. Additionally, over-learned stimulus-response links that serve as the basis for automaticity may often compete with intentions, as has been studied extensively in the cognitive control literature. There will often be a need to exert control to favor the production of a task-relevant response.

In short, an approach to control must be developed and implemented that serves conflict monitoring, suppression of unwanted automatic responses and distracters, scheduling and switching between tasks as well as maintenance of information to support intentional action. Finally and perhaps most importantly, practical reasoning must be recast in terms of intentional actions if we are ever to cleanly make a distinction between an agent making considered choices with respect to its background knowledge, norms, and values, and cases where the appearance of choice is generated by extra-intentional processes that the agent has no part in actively guiding, as was true of Smith's erratic driving. The formulation of reasoning as a form of intentional action is both novel and challenging. Part of the challenge lies in the fact that the best-developed theory of human reasoning is expressed as a model-based dual-process theory and disavows the idea that rule-like content is involved in naive reasoning whatsoever (Johnson-Laird, 2006). A computational rendering of the theory has been produced (Khemlani and Johnson-Laird, 2021), but content involving norms and other complex semantic content has yet to be explored in the computational framework. Alternatively, many approaches to automated reasoning with rules or graphical models exist, but they are disconnected from the literature on human reasoning and on cognitive architecture, leaving the challenge of finding intersection points with attention, memory, and other supporting cognitive processes. Much work remains to be done to produce initial results that offer the generality of AI approaches to reasoning with the empirical constraints provided by existing psychological accounts.

In the near future, ARCADIA will be extended to embody ideas from new psychological work on the expected value of control (EVC) (Shenhav *et al.*, 2013). Within the EVC framework, control configurations are selected by virtue of their expected value. The computations fold in information about costs, including lost opportunity since selecting one particular control configuration rules out the selection of others. When value is defined as a relative contribution to achieving overall task goals EVC gives a rational analysis of controlled behavior. To date, EVC has been primarily applied to psychological lab tasks, where it is often straightforward to formalize notions of control, cost, and value. It remains to be seen if costs and values are able to be represented simply in real-world decision problems. In the case of ARCADIA, control configurations will encompass shifts of attention realized by increased maintenance of the task set, engagement over time with particular focal elements to maximize interaction with components through the broadcast process, disengagement with performance-degrading focal elements, and reordering of attentional priorities. Failures of control, such as Jones's, can be explained via momentary over- and under-estimation of costs and values in the EVC process, whereas success can be explained by proper estimation. Although existing work on EVC hints at costs and values being learned and not necessarily as content for outside reasoning or decision-making, this may only reflect aspects of the tasks it has been applied to rather than a set of hard constraints on its use. Instantiated in ARCADIA, we conjecture that the EVC framework can be extended and scaled to more complex behavior.

6. Final Thoughts

In many ways, this chapter has been a strange discussion given the objective of detailing the relationship between attention, consciousness, and agency. We have generously covered attention and some aspects of agency, particularly related to intentions, while remaining more circumspect in theorizing about consciousness. Part of our hesitance is due to the ever-expanding bounty of divergent views by various leading researchers on what the fundamental nature and role of consciousness even is. Even if we leave aside the discussion about the hard problem of consciousness, debates still rage among psychologists, neuroscientists, and philosophers about whether consciousness is a function of the capacity to have higher-order thoughts, whether it is the contents of a global workspace, a function of information integration as constrained by neural architecture, or even whether it can be present in the absence of attention. Although ARCADIA is in the vicinity of a global workspace view of cognitive architecture and the analyses of the two cases offered in this chapter take advantage of this fact, it seems plausible enough that alternative analyses could be offered conditioned on other models of consciousness. However, the critical question shared across all of these possible models is whether consciousness depends on attention. This issue is hotly debated among consciousness researchers and nothing presented here is designed to make progress on resolving it. However, it is useful to point to places where an answer to the question has broad implications and discussions of agency and intentional action certainly seem to provide a backdrop for asking them.

Turning to the computational, explorations at the intersection of consciousness, agency, and intentional action have had new life given to them by surging interest among researchers in human-machine interaction, autonomous systems, and the burgeoning interest in machine and AI ethics in recent years. Historically, AI has either been seen as a niche pursuit among denizens of the ivory tower or as a set of fairly banal computational prostheses. Until the recent present, few thought of how ubiquitous these technologies would become in society or the potential harms they might cause. Of course, where there are harm done by systems that are perceived to be intelligent or by human-machine teams, there arises the question of responsibility. Humans see each other as conscious agents with foresight and the capability for control, and these features often underwrite intuitions about responsibility. Whether humans perceive machines as potential targets for responsibility judgments, and if so, which features bear on such judgments are still open questions sorely in need of further study. But if machines are to one day be in some sense responsible agents it is important to computationally explore relationships between the aspects of consciousness and attention that support agency. Here we have argued that attention plays a central role in mediating intentional action and have given an early demonstration of the ARCADIA attention-centric cognitive system. It is fitting that the part of this chapter devoted to limitations and future directions is quite large, for there is much to be done still.

References

Baars, B. J. (1997). In the theatre of consciousness. global workspace theory, a rigorous scientific theory of consciousness, *Journal of Consciousness Studies* **4**, pp. 292–309.

Bello, P. and Bridewell, W. (2017). There is no agency without attention, *AI Magazine* **38**, 4.

Bello, P., Lovett, A., Briggs, G., and O'Neill, K. (2018). An attention-driven computational model of human causal reasoning, in *Proceedings of the 40th Annual Meeting of the Cognitive Science Society*, pp. 1353–1358.

Bridewell, W. and Bello, P. (2016a). A theory of attention for cognitive systems, in *Fourth Annual Conference on Advances in Cognitive Systems*, Vol. 4 (Evanston, IL), pp. 1–16.

Bridewell, W. and Bello, P. F. (2015). Incremental object perception in an attention-driven cognitive architecture, *Proceedings of the 37th Annual Meeting of the Cognitive Science Society*, pp. 279–284.

Bridewell, W. and Bello, P. F. (2016b). Inattentional blindness in a coupled perceptual–cognitive system, in *Proceedings of the 38th Annual Meeting of the Cognitive Science Society* (Philadelphia, PA), pp. 2573–2578.

Bridewell, W., Wasylyshyn, C., and Bello, P. (2018). Towards an attention-driven model of task switching, *Advances in Cognitive Systems* **6**, pp. 1–6.

Briggs, G., Bridewell, W., and Bello, P. F. (2017). A computational model of the role of attention in subitizing and enumeration, in *Proceedings of the 39th Annual Meeting of the Cognitive Science Society*, pp. 1672–1677.

Broughton, R., Billings, R., Cartwright, R., Doucette, D., Edmeads, J., Edwardh, M., Ervin, F., Orchard, B., Hill, R., and Turrell, G. (1994). Homicidal somnambulism: a case report. *Sleep* **17**, 3, pp. 253–64.

Chisholm, R. M. (1964). The descriptive element in the concept of action, *The Journal of Philosophy* **61**, 20, p. 613, doi:10.2307/2023443.

Cohen, P. R. and Levesque, H. J. (1990). Intention is choice with commitment, *Artificial Intelligence* **42**, 2-3, pp. 213–261, doi:10.1016/0004-3702(90)90055-5.

Dehaene, S., Changeux, J. P., Naccache, L., Sackur, J., and Sergent, C. (2006). Conscious, preconscious, and subliminal processing: a testable taxonomy, *Trends in Cognitive Sciences* **10**, 5, pp. 204–211, doi:10.1016/j.tics.2006.03.007.

Duncan, J., Williams, P., Nimmo-Smith, I., and Brown, I. (1993). The control of skilled behavior: Learning, intelligence, and distraction, doi:10.7551/mitpress/1477.003.0022.

Johnson-Laird, P. N. (2006). *How We Reason* (Oxford University Press, New York, NY, US).

Khemlani, S. and Johnson-Laird, P. N. (2021). Reasoning about properties: A computational theory. *Psychological Review*, doi:10.1037/rev0000240.

Levy, N. (2014). *Consciousness and Moral Responsibility* (Oxford University Press).

Lovett, A., Bridewell, W., and Bello, P. (2019). Selection enables enhancement: An integrated model of object tracking, *Journal of vision* **19**, 14, p. 23, doi:10.1167/19.14.23.

O'Neill, K., Bridewell, W., and Bello, P. (2018). Time-based resource sharing in arcadia, in *Proceedings of the 40th Annual Meeting of the Cognitive Science Society*, pp. 828–833.

Padgham, L. and Winikoff, M. (2005). *Developing Intelligent Agent Systems: A Practical Guide* (Wiley).

Rao, A. S. and Georgeff, M. P. (1991). Modeling rational agents within a BDI-architecture, in *Proceedings of the Second International Conference on Principles of Knowledge Representation and Reasoning*, pp. 473–484.

Russell, S. J. and Norvig, P. (2010). *Artificial Intelligence: A Modern Approach* (Pearson Education, Inc.).

Shenhav, A., Botvinick, M. M., and Cohen, J. D. (2013). The expected value of control: an integrative theory of anterior cingulate cortex function, *Neuron* **79**, 2, pp. 217–240, doi:10.1016/j.neuron.2013.07.007.

Watzl, S. (2017). *Structuring Mind. The Nature of Attention and How it Shapes Consciousness* (Oxford University Press).

Wu, W. (2016). Experts and deviants: The story of agentive control, *Philosophy and Phenomenological Research* **93**, 1, pp. 101–126, doi:10.1111/phpr.12170.

Chapter 8

Artificial Conscious Intelligence:
Why Machine Consciousness Matters to AI

†James A. Reggia, ‡Garrett E. Katz, §Gregory P. Davis

†*Dept. of Computer Science, University of Maryland,*
College Park, MD 20742 USA
reggia@umd.edu
‡*Dept. of Elec. Engr. and Comp. Sci., Syracuse University,*
Syracuse, NY 13244 USA
gkatz01@syr.edu
§*Dept. of Physiology and Biophysics, Weill Cornell Medicine,*
New York, NY 10065 USA
gpd4001@med.cornell.edu

The fields of artificial intelligence (AI) and artificial consciousness (AC) have largely developed separately, with different goals and criteria for success and with only a minimal exchange of ideas. In this chapter, we consider the question of how concepts developed in AC research might contribute to more effective future AI systems. We first briefly discuss several past hypotheses about the function(s) of human consciousness, and present our own hypothesis that short-term working memory and very rapid learning should be a central concern in such matters. We describe our recent efforts to explore this hypothesis computationally and to identify associated computational correlates of consciousness. We then present ideas about how integrating concepts from AC into AI systems to develop an artificial conscious intelligence (ACI) could both produce more effective AI technology and contribute to a deeper scientific understanding of the fundamental nature of consciousness and intelligence.

Keywords: artificial consciousness; artificial intelligence; computational correlates of consciousness; computational explanatory gap; neural virtual machine; working memory

1. Machine Intelligence versus Machine Consciousness

In his foundational paper titled "Computing Machinery and Intelligence", Alan Turing [1950] did the emerging field of artificial intelligence (AI) a great service. At the time, just a few years after the construction of the first electronic digital computers, there was a lot of discussion about whether or not a machine could think or have a mind. Turing found questions like these to be "too meaningless to deserve discussion", and instead proposed what we today call the Turing Test as a criterion for whether machines "can think". This test has often been taken as a criterion for machine intelligence in AI. While this specific criterion has faced well-deserved criticism during contemporary times, Turing's basic notion that we should assess such issues based on a machine's *behavior,* rather than on vaguely defined concepts such as the existence of an underlying mind, was liberating and persists as the dominant paradigm in AI to this day. In a very real sense Turing's approach made research into AI respectable. This is because the idea of machine intelligence only refers to whether a machine can exhibit intelligent behavior, and it does not represent a claim to having created a machine that has subjective mental experiences, can think, or have a mind. Avoiding these latter difficult issues has, to a great extent, enabled the successful pursuit of AI as a technology.

On the other hand, an additional consequence of this dominant viewpoint is that, with very few exceptions, it has largely side-lined work in AI on challenging issues surrounding the possibility of an artificial mind or a conscious machine. Many AI researchers find such issues to be uninteresting or insufficiently defined to be of any relevance to AI [Bringsjord, 2007; McDermott, 2007]. As a result, the field of artificial consciousness (AC) has largely developed outside of mainstream AI [Reggia, 2013]. In our opinion this is regrettable because there is substantial room for synergistic work in these two fields.

We have previously considered the issue of how work in AI might contribute to advancing AC [Reggia *et al.*, 2014, 2017]. Our central point in this regard is that a *computational explanatory gap* currently limits our ability to advance work in AC. The computational explanatory gap is our lack of understanding of how consciously accessible high-level cognitive information processing can be mapped onto low-level neural computations.

The computational explanatory gap is a purely computational issue and not a mind-brain problem — it is a gap in our understanding of how cognitive algorithms (executive control, goal-directed problem solving, planning, etc.) can be mapped into the sub-symbolic computations supported by neural networks that use a distributed representation of information. This issue is clearly relevant to AI in general, and encouragingly increasing attention is being paid to it in studying "programmable neural networks", for example [Devlin *et al.*, 2017; Graves *et al.*, 2016; Katz *et al.*, 2019]. The computational explanatory gap also makes cognitively oriented models in AI much more relevant to AC than is often recognized, especially given the philosophical concept of cognitive phenomenology.[a]

Having previously considered how work in AI may contribute to AC, here we address the opposite question: How might concepts developed via work in AC and consciousness studies in general enhance the functionality of AI systems, making them more effective? To answer this question, we first summarize some past ideas about what the function of consciousness is in people (Sec. 2). We next present our own hypothesis that working memory and very fast learning/unlearning of its contents should be a central concern in such matters (Sec. 3), summarizing the results of some work we have done studying this issue (Sec. 4), and describing our most recent efforts to increase the power and effectiveness of working memory models (Sec. 5). With these considerations about the function of biological consciousness and their relations to working memory in hand, we then return to the question of how work in AC is relevant to the development of future AI systems (Sec. 6), and we provide a summary of our conclusions on these matters (Sec. 7). In particular, we suggest that integrating concepts from AC into AI systems to develop an artificial conscious intelligence (ACI) could both produce more effective AI technology and contribute to a deeper scientific understanding of the fundamental nature of consciousness and intelligence.

---

[a]Cognitive phenomenology asserts that parts of our cognitive processes are consciously accessible above and beyond their sensory representations [Bayne and Montague, 2011]. Our other points about the significance of the computational explanatory gap hold regardless of the validity of cognitive phenomenology.

2. What is the Function of Human Consciousness?

We approach the question of what consciousness might contribute to AI systems by first asking what its function is in human cognition, and then exploring whether such functionality might provide/improve similar, currently absent/limited functionality in machine intelligence. This of course presumes that consciousness does have a biological function, an assumption that we make here along with the assumption that consciousness has a physical basis. While such assumptions are controversial, with some arguing that consciousness is just an epiphenomenon, we note that the evolution of consciousness in at least humans and some animal species supports the idea that it contributes to survivability and reproductive fitness, and we explore the consequences.

There is no shortage of past hypotheses concerning the function(s) of human consciousness. Here we take asserting that something is a *function* of consciousness to implicitly indicate that a causal relationship is involved: that consciousness causes and is in large part necessary for that function. Many AC investigators have hesitated to claim that consciousness causes or is caused by various factors, and they have instead focused on considering neural or computational *correlates* of consciousness. A neural correlate of consciousness is a minimal neurobiological state whose presence is sufficient for the occurrence of a corresponding state of consciousness [Metzinger, 2000]. A computational correlate of consciousness is a minimal computational mechanism that is specifically associated with conscious aspects of cognition but not with unconscious aspects [Cleeremans, 2005; Reggia *et al.*, 2014]. Being a function of consciousness generally implies being a correlate of consciousness, but not vice versa.[b] With this understanding we now give a non-exhaustive listing of functions of consciousness previously proposed in the literature, ordered arbitrarily:

- global access to & integration of information [Baars, 1997; Tononi, 2008]

---

[b]A neural/computational correlate of consciousness could be something caused by consciousness, something that causes consciousness, or neither (for example, there might be a separate underlying cause of both consciousness and the correlate).

- symbol grounding [Chella, 2008; Kuipers, 2008; Haikonen, 2019]
- high-level symbolic cognition [Sun, Franklin, 2007; Pasquali *et al.*, 2010]
- supporting executive functions [Shanon, 1998; Rosenthal, 2008]
- error detection and correction [Baars, 1997; Taylor, 2007]
- novelty detection and generation [Baars, 1997; Mudrik *et al.*, 2012]
- self-awareness [Perlis, 1997; Holland, 2007; Chella, 2008; Takeno, 2013]
- source of intrinsic motivation [DeLancey, 1996; Sanz et al., 2012]
- evoking/informing volitional actions [Earl, 2014; Pierson, Trout, 2017]
- attention mechanisms, control [Taylor, 2007; Haikonen, 2019]

The large number of these past hypotheses is remarkable, but it is consistent with the sizable number of theories about the nature of consciousness in general [Katz, 2013]. This is ameliorated somewhat by the fact that these hypotheses are generally not mutually exclusive or independent (e.g., symbol grounding and inference [Brody *et al.*, 2016]), and it could be that consciousness has multiple functions. Our point here is that, at the present time, there is no clear consensus on an identifiable function of consciousness that provides an adaptive advantage. For example, several of the potential functions of consciousness listed above have been criticized on various grounds due to inconsistency with empirical data or theoretical considerations [Manzotti, 2012; Mudrick *et al.*, 2012; Rosenthal, 2008; Seth, 2009].

3. Memory, Learning and Consciousness

Contemporary AI recognizes that intelligent agents can be composed of multiple functional components, some of which deal with the *processing* of information (reflexive condition-action rules, symbolic reasoning, executive decision making, taking actions, etc.) and some of which deal with the *memory and learning* of information [Russell and Norvig, 2010]. From this AI perspective, it is striking that the diverse list of previously proposed functions of consciousness in the preceding section generally have one thing in common: They largely deal with some facet of the *processing* of information (its integration, manipulation, use for inference or decision making, etc.). In contrast, here we propose an alternative, complementary possibility that the adaptive function of human consciousness is to be found in its contribution to *memory and learning* rather than to the subsequent processing of that information. Specifically, *we hypothesize that the fundamental function of consciousness and its contribution to intelligence will most likely be found in its role in supporting short-term working memory and its associated learning and control mechanisms.*

Psychologists distinguish different memory systems in explaining various neuroscientific and behavioral data [Squire and Dede, 2015]. Human memory at the top level is often characterized in terms of long-term memory versus short-term memory. Long-term memory is then sub-divided into distinguishable types, such as semantic, episodic, and procedural memory, and we do not consider these further in this chapter. Short-term memory is also sub-divided into types, one of which is *working memory* and that serves as our focus here (we do not consider other types of memory in this work, such as short term visual iconic memory [Pratte, 2018]). Working memory stores recently experienced information, typically for a period of seconds to minutes, that is being used in problem solving or other cognitive activities. In contrast to long-term memory with its enormous storage capacity, short-term working memory is characterized by a very limited capacity, and it is able to retain just a few independent items at any one time [Cowan *et al.*, 2005; Kruijne *et al.*, 2021].

Why focus on working memory as a function of consciousness? One reason is that working memory is widely recognized in philosophy and psychology to involve conscious, reportable cognitive activity [Baars and Franklin, 2003; Baddeley, 2012; Carruthers, 2015; Persuh *et al.*, 2018]. Our view of this relationship is that what psychologists refer to as "working memory" is mostly the same as what some philosophers would characterize as the state of a conscious mind. For information to be consciously accessible and reportable essentially requires that information to be actively represented in working memory. Whether there are also things that are in working memory that are not conscious, or there are things that are conscious but not in working memory, are open questions at present.

Another reason for focusing on working memory is that it is a fundamental underlying element of cognition that provides a unifying perspective for the multiple possible functions of consciousness that have been proposed in the past (listed in the previous Section). For example, the neurobiological mechanisms that underlie working memory appear to be fairly widespread throughout cerebral cortex [Lara and Wallis, 2015], consistent with the hypothesis that consciousness supports global access to and integration of information. The representation of symbolic information in working memory supports the importance of symbol processing and grounding in human consciousness. The top-down, goal-directed control of working memory that distinguishes it from low-level sensorimotor processes is consistent with past proposals that high-level cognition, executive functions, and attention mechanisms are all key aspects of conscious mind. In other words, what makes working memory "working" is that its contents are actively manipulated by cognitive processes: it is at the intersection of algorithms and data structures. Working memory may turn out to be a common underlying factor in all of these previously proposed functions of consciousness since it is such a foundational aspect of human cognition.

4. Working Memory and the Computational Correlates of Consciousness

Can computational models of working memory suggest any specific computational correlates of consciousness that might ultimately be used to enhance AI systems? We have recently been examining this issue. There are many different ways of implementing a given piece of functionality in simulated neural networks. Our initial work focused on application-specific models based on standard psychological tests of working memory such as the n-back task [Sylvester *et al.*, 2013] and on solving problems involving card matching tasks [Sylvester and Reggia, 2016]. Recently we greatly generalized our computational models of working memory in the context of developing a *neural virtual machine* (NVM) that is capable of universal computation [Katz *et al.*, 2019].[c] The NVM is a purely neurocomputational, application-independent software environment that allows one to instantiate cognitive-level algorithms in neural networks with rate-coding (non-spiking) neurons. Such algorithms are currently readily implemented via traditional symbolic AI methods, but much less so via existing programmable neural network methods. Importantly, the NVM's modeled knowledge and cognitive processes are acquired through a learning process and represented by distributed patterns of activity over an underlying neural substrate. From a user's perspective, to model a cognitive process using the NVM one writes an assembly language level program for a virtual machine that is emulated by the NVM (see Fig. 1(a)). However, in actuality, the NVM converts that given program into a region-and-pathway system of recurrently connected neural networks that perform the indicated computations on distributed activity patterns representing symbols, based on the network's dynamics and synaptic weight changes. In short, the NVM can be viewed as a step towards bridging the computational explanatory gap: unlike hybrid systems it is purely neurocomputational. A detailed description of the NVM with a link to an open-source implementation is available [Katz *et al.*, 2019].

How can we use these models of working memory to develop a better understanding of consciousness? Much past work in AC considering computational correlates has started from the premise that some underlying mechanism is a key aspect of consciousness (global processing, attention, self-modeling, etc.) and then explored the implications of that premise via computational modeling. In contrast, our recent work has taken the opposite approach: Start with a model of working memory and ask what core, distinguishing neurocomputational mechanisms are needed to implement that model. The idea is that such distinguishing mechanisms suggest new candidates for computational

---

[c]The NVM is only capable of universal computation in the limit as the number of neurons goes to infinity. While none of our neurocomputational working memory models described here are intended to capture biologically-realistic neural circuitry, they incorporate separate modules for working memory proper (these store ongoing problem solving information, which is believed to be widely distributed across human cerebral cortex) and other modules for representing executive-level control of working memory functionality (these store action sequences, and are most closely related to human prefrontal cortex) [Lara and Wallis, 2015].

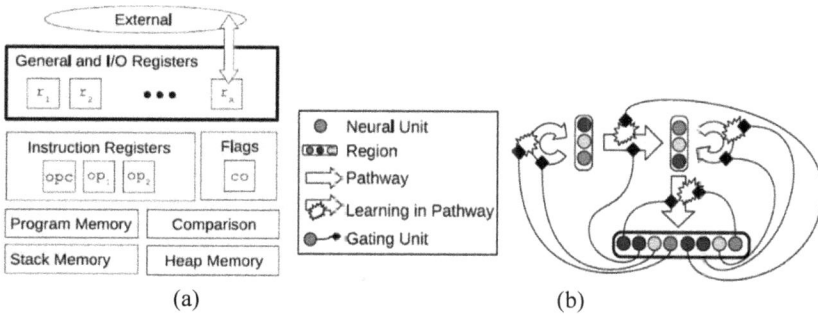

(a)   (b)

Fig. 1. (a) The virtual machine emulated by the NVM. (b) A toy example of an underlying recurrent neural network supported by the NVM that uses gating connections to modulate activation and learning in other pathways. Each block arrow represents many individual connections (not shown) that are being gated.

correlates of consciousness, based on the fact that working memory is tightly associated with conscious mind. Such correlates might be useful in AI systems. This approach has led us to suggest three new correlates, all of which are incorporated into the NVM's implementation of working memory, as follows.

The first potential computational correlate of consciousness that we identified is learned *itinerant attractor sequences*, i.e., sequences of learned attractor states of the underlying recurrent neural network's activity, where each sequence element represents a cognitive state of working memory. Each learned attractor state corresponds to an action or "instruction" that is currently active in working memory as a task is being performed. In contrast to previous proposals that individual attractor states or activity trajectories in general may be computational correlates of consciousness, we specifically mean that (i) the trajectory is composed of a sequence of attractors, (ii) this sequence contributes to control of agent behavior and working memory processes, (iii) it involves learned states rather than pre-wired genetically-determined circuitry, and (iv) it involves cognitive states used in high-level problem solving and reasoning. Such sequences can represent not only arbitrary procedures, but also arbitrary data structures in general such as lists or trees. This computational correlate supports past suggestions that the functions of consciousness include symbol processing (each attractor can be viewed as representing a symbol in working memory) and error detection/correction (the transitions between working memory states need only be approximate as the system's dynamics will correct for errors by converging on the nearest attractor state).

The second possible computational correlate of consciousness suggested by our modeling work is the *top-down gating* of working memory by which high-level cognitive processes control what is stored, manipulated and learned by working memory. For example, as illustrated by the gating units in Fig. 1(b), the underlying neural networks that the NVM uses to implement given algorithms make heavy use of multiplicative gating to turn on/off the flow of information through a network's pathways and to enable/disable learning on network connections. Each action/instruction in a procedural attractor sequence is performed by using multiple coordinated gating operations. We postulate that these gating operations, driven by the sequences of attractor states in the executive component of our working memory models, represent consciously reportable cognitive activities in working memory, and for that reason we take them to be possible computational correlates of consciousness that may contribute to a sense of agency and mental causation. This computational correlate supports past suggestions that the functions of consciousness include top-down executive processes, attention mechanisms, and the evoking/controlling of actions.

Finally, the third computational correlate is *very fast weight changes* that provide for immediate, simultaneous one-step learning and unlearning in working memory. Human short-term working memory is remarkable in its ability to reliably learn new information immediately from just a single presentation of that information. For example, if one is verbally told "add 16 to 17", the numbers involved are immediately retained in working memory as the computation is done — there is generally no need for a person to hear the problem stated several times to learn what the problem is. Such learning is very different from what is done in many neural network learning systems, including those based on gradient descent methods that require numerous iterative presentations of material to be learned. Our recent modeling work with the NVM implements very fast additions/deletions to working memory contents via synaptic weight changes that involve simultaneously using (1) one-step Hebbian learning to retain new information in working memory, and (2) one-step anti-Hebbian unlearning that actively removes old information that is no longer needed in working memory [Katz *et al.*, 2019]. This fast *store-erase learning rule* introduced in the NVM is responsible for the NVM's ability to control dynamically what is retained and what is removed from working memory during problem solving. Further, it has proven effective for both representing information about the state of a problem being solved (temporally symmetric weight changes) and information about the behavioral action sequences that control problem-solving (temporally asymmetric weight changes). This third computational correlate suggests, for the first time to our knowledge, that the very fast learning/unlearning of information in working memory may be an important function of consciousness.

5. Increasing the Power of Working Memory

The importance for understanding consciousness that we have assigned to working memory in general, and to the potential computational correlates of consciousness described in the previous section, naturally give rise to further questions. For example, how can we better understand working memory computationally, and how can we increase its effectiveness? To address these questions, we have most recently used computer simulations to examine three issues: why the one-step store-erase learning rule has proven to be so effective, whether our basic model of working memory can be extended to handle compositionality, and whether the model can be extended to learn procedural information via reinforcement learning. We briefly address each of these points in the following.

5.1. *Working memory and catastrophic forgetting*

The topic of *catastrophic forgetting* has become a major issue in many neural network models of long-term memory undergoing continual or life-long learning [Chen and Liu, 2018; Parisi *et al.*, 2019; Xiong *et al.*, 2020]. As the amount of information being learned increases over time, a neural network eventually experiences "memory overload". New items cannot be added sequentially to memory without impairing a network's ability to faithfully recall earlier stored items. This issue is even more critical in neural models of short-term working memory due to their very limited memory capacity and their transient retention of information that requires a mechanism not only for storing items in memory, but also for removing/erasing items to make room for new information. Catastrophic forgetting in working memory models is especially malignant in that it obliterates not only old previously stored items but also the ability to store new items. There can be a complete collapse of learning. The critical issue is to selectively maximize the *persistent memory capacity* (the number of recently stored items that can be retrieved correctly) over an infinite time horizon, while selectively forgetting older memory patterns that were stored in working memory.

As noted above, in developing a neural virtual machine (NVM), we introduced a new *store-erase learning rule* for working memory [Katz et al., 2019]. In a single time-step, this learning rule both stores a new associative memory pattern using Hebbian learning and simultaneously erases previously stored associative information using anti-Hebbian learning. However, this work used the store-erase learning rule for only a single continuous activation function, and it did not explicitly consider the issue of catastrophic forgetting. We thus recently extended this past work by systematically examining the extent to which the store-erase learning rule can ameliorate catastrophic forgetting over long time horizons in models of short-term working memory that are using discrete-valued activation functions [Reggia et al., 2021].

For example, in one computational experiment a single layer neural network of linear threshold units having 100 input units and 100 output units was trained to associate M input patterns with corresponding output patterns. When basic Hebbian learning was used alone (Fig. 2, top panel), catastrophic forgetting occurred: As the number M of associations stored increased, the network's persistent memory capacity dropped to zero. In contrast, when a store-erase learning rule was used instead, a persistent memory capacity of about 12 patterns was achieved as M increased over time (Fig. 2, bottom panel). In general, we found that store-erase learning, especially when combined with weight decay, provides the optimal avoidance of catastrophic forgetting while still selectively retaining the most recently stored memories.

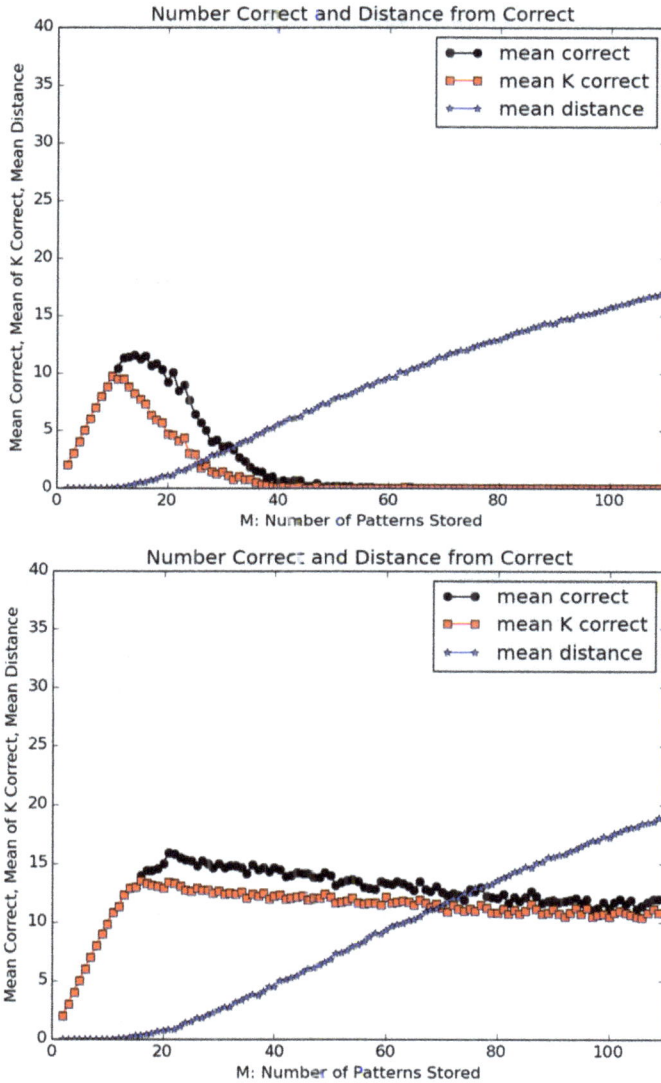

Fig. 2. Simulations of working memory involving heteroassociative learning using Hebbian learning (top panel) versus store-erase learning (bottom panel). 100 input units, 100 output units, linear threshold units. Correctly learned associations (black), K most recent learned associations (red), and Hamming distances of output patterns from stored ones (blue).

This result, while obtained only for the case of short-term memory here, may have much broader implications for neural AI systems. Weight decay is widely used in contemporary deep convolution networks as a regularization technique to improve generalization, but it is a kind of blind unlearning of older information. Based on our experiences with these recent computational experiments, we conjecture that anti-Hebbian learning as used with the store-erase rule, by providing a more targeted weight erasure than weight decay, might help prevent catastrophic forgetting in such neural systems during learning if appropriate criteria can be identified for when older stored information is no longer needed.

5.2.   *Working memory and compositionality*

Compositionality refers to the ability of an intelligent system to construct representations out of reusable parts [Fodor and Pylyshyn, 1988; Nefdt, 2020]. Substantial evidence indicates that working memory is compositionally structured, and that its organization is strongly related to learned cognitive processes. Compositional reasoning is considered crucial in domains as diverse as language comprehension, behavioral planning and imitation, visual perception, and concept learning. Thus, understanding how the brain manipulates compositional structures in working memory may shed significant light on the neural basis of cognition in general.

Compositionality is readily achieved in symbolic AI systems, but it is much more challenging for sub-symbolic systems such as artificial neural networks. Recent advances in machine learning, particularly in natural language processing, have reignited a long-standing controversy over whether neural networks can represent compositional structures [Lake *et al.*, 2017]. Still, empirical studies show that state-of-the-art neural networks struggle to learn systematic rules for compositional reasoning, limiting their ability to generalize beyond training data [Lake and Baroni, 2018; Hupkes *et al.*, 2020]. Why is this?

Working memory in artificial neural networks is typically based on persistent maintenance of activity patterns, often in specialized memory arrays that are accessed with neural attention [Graves *et al.*, 2016]. Such memory arrays are considered highly implausible from a biological perspective. Furthermore, substantial evidence indicates that activity-silent mechanisms such as rapid synaptic plasticity (i.e., very fast weight changes) play a critical role in working memory [Manohar *et al.*, 2019; Rose *et al.*, 2016]. Can we improve the compositionality of artificial neural networks with a neural working memory system based on our three identified correlates?

To answer this question, we recently developed *attractor graph networks*, which represent compositional data structures as systems of attractor states (vertices of the attractor graphs) linked with learned transitions (edges) [Davis *et al.*, 2021]. Whereas states in attractor sequences are limited to a single successor, states in attractor graphs may have several outgoing transitions to other states, each with a distinct label. During traversal through an attractor graph, transitions are selected by a sequence of multiplicative input patterns that represent edge labels. Each input pattern is a top-down gating signal that selects a subset of the neural population to participate in transition dynamics and learning. Thus, different transitions from the same state are stored in distinct but overlapping subsets of the hetero-associative weight matrix. We refer to this as *functional branching*, because branches in the attractor graph depend on patterns of activity and are not stored in distinct weight matrices.

We hypothesized that compositional data structures could be represented as attractor graphs, and that they could be efficiently constructed and accessed via algorithmic procedures that control top-down gating over time. We tested these hypotheses with several computational experiments [Davis *et al.*, 2021]. We found that attractor graph networks can reliably learn graphs with large numbers of transitions (edges) and with very high branching factors (node out-degrees) using the fast store-erase learning rule (for example, see Fig. 3). Further, the store-erase learning rule permits rapid reorganization of attractor graphs, making the network an effective model of reusable working memory. We showed that data structures such as lists, trees, and associative arrays can be encoded as attractor graphs by imposing constraints on their organization (e.g., lists are chains of states composed of edges with a shared list-specific label). These data structures can be constructed, manipulated, and accessed via programmatic top-down control. To demonstrate this, we designed a programmable neural network with an attractor graph network as a memory module, and we trained it to perform a hierarchical planning task. In this task, the network is required to transform a sequence of abstract behaviors into a sequence of concrete actions according to learned decomposition rules with environmental contingencies (e.g., opening a door requires different actions depending on what type of door it is). To do so, the network constructed a tree in working memory representing the behavioral plan by recursively decomposing abstract actions, and then traversed it to retrieve the leaf nodes representing concrete actions. We found that the network could reliably complete the task with a sufficiently sized attractor graph network.

Fig. 3. Illustration of the practical implications of using the store-erase learning rule rather than basic Hebbian learning for storing data structures as attractor graph networks [Davis *et al.*, 2021]. With Hebbian learning, as a sequence of trees is stored (each tree being one trial) catastrophic forgetting occurs — the accuracy of perfect recall drops to essentially zero. In contrast, with the same network the store-erase rule retains the ability to store and retrieve the same sequence of trees. Analogous results based on similarity metrics can be found in [Davis *et al.*, 2021].

Our results confirm that neural systems with compositional working memory can be developed using the three computational correlates of consciousness that we have identified. Furthermore, they show how activity-dependent gating of neural populations increases the expressivity of neural working memory, permitting representation of complex data structures that can be constructed and accessed via coordinated top-down gating procedures. From a computational perspective, this suggests that some cognitive processes can be understood as algorithmic procedures that reorganize data structures in working memory via top-down control of learning. In turn, we might expect that cognitive procedures are organized compositionally in working memory. We are currently investigating this with neural systems that use attractor graphs to represent structured programs written in high-level programming languages.

5.3. *Working memory and reinforcement learning*

The NVM we described previously is a purely neural system that can emulate source code provided by a human. At the core of this emulation mechanism, a special gating layer transits through sequences of activity patterns, which selectively allow or inhibit associative learning and recall in the other working memory layers. This provides a model of storing and executing procedural knowledge in working memory, when that procedural knowledge is provided explicitly. However, it does not address how conscious artificial agents might acquire procedural knowledge on their own, based on their experiences while interacting with the environment.

In subsequent research, we studied reinforcement learning (RL) in the NVM without explicit human-provided source code [Katz *et al.*, 2020; Lin *et al.*, 2021]. In this work, the NVM was trained to perform various algorithmic list processing tasks, but it was not explicitly told what the tasks were, or how to solve them. Instead, it was given many training examples of unprocessed lists. When it processed them correctly, it received a reward. Over the course of training, the NVM gradually discovered its own strategies for how to gate its working memory, in such a way that maximized average reward on a given task.

For example, one task was reversing the order of elements in a list. In each training episode, a random list of symbols was fed into the NVM, one at a time. If the output sequence of the NVM contained the same random symbols, but in reverse order, then it received a numeric reward of +1. Otherwise, it received zero reward. The precise timing of the output symbols was not constrained, so long as the output sequence contained the correct symbols in the correct order. At time-steps where the NVM was not yet ready to produce the next list symbol, it was allowed to produce one or more occurrences of a special "padding" symbol. This afforded the NVM more flexibility in determining its own solution strategy, which is why we used reinforcement as opposed to supervised learning.

We found empirically that the NVM could learn several list processing tasks in this way, using classic vanilla policy gradient optimization [Williams, 1992] to train the weights of the NVM's gating region. For example, across 30 independent repetitions of the experiment, the NVM consistently learned how to identify the maximum element of a list or filter out all list elements below a cut-off (see Fig. 4).

The NVM also learned to reverse the order of some short lists, although the result was less consistent across repetitions of the experiment. However, NVM learning could be further improved by using a more modern RL technique, namely proximal policy optimization, and more fine-grained reward signals that gave partial credit for outputs that were partially correct. These improvements enabled the NVM to more consistently reverse longer lists, and facilitated learning of some new tasks, including list sorting with short lists.

The tight integration of reinforcement learning and working memory in this trainable NVM model is concordant with neurobiological research [Collins and Frank, 2012] which suggests that in humans, working memory plays an important role in reward-based instrumental learning. In our work so far, the gating layer learns how to manipulate working memory to solve a given task. The neuroscientific theories posit that beyond that, working memory is recruited as part of the learning process itself: hypotheses about how action sequences lead to rewards (i.e., model based RL) are also stored in working memory, and the agent tests those hypotheses in order to learn something new. In humans, this recruitment of working memory during RL may be an important computational correlate of consciousness with substantial utility for AI system design. Our future work with the NVM and RL will explore how working memory could participate in the learning process in this way.

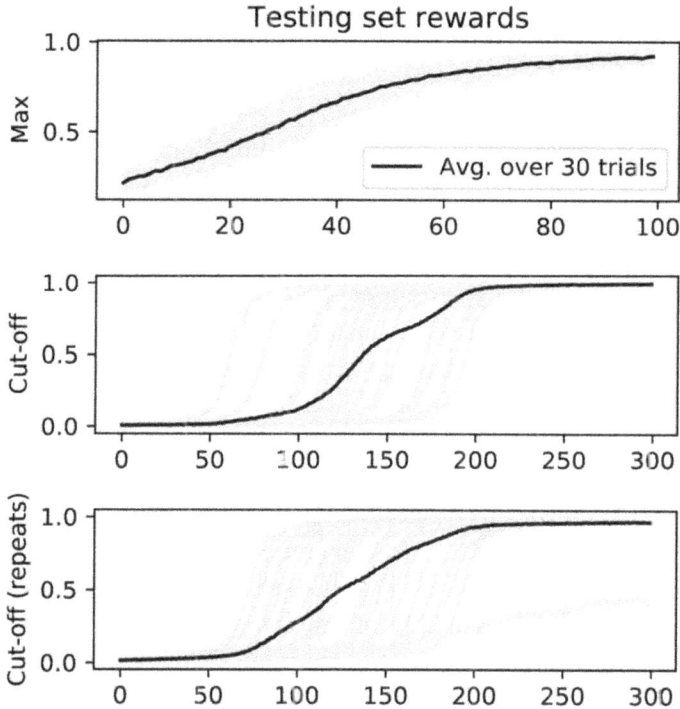

Fig. 4. Average reward over the course of training on a test set of held-out examples, for three list processing tasks. Training iteration is shown on the x-axis. Each gray curve is an independent repetition of the experiment; black curves are averages across all repetitions.

6. ACI = AI + AC

We will refer to AI systems that incorporate concepts from AC as *artificial conscious intelligence* (ACI).[d] Put simply, ACI = AI + AC. Having considered above what the function(s) of consciousness might be, we now

---

[d]This is analogous to distinguishing artificial general intelligence (AGI) systems that study general purpose AI from the more common practice of creating application-specific AI systems.

return to our central question: How might work done in AC enhance the functionality of future AI systems? There are at least two distinct answers to this question about ACI depending on whether one considers simulated or instantiated machine consciousness.[e] By *simulated consciousness*, we mean simulations that attempt to capture some aspect of consciousness or its neural or behavioral correlates in a computational model. Most work in AC falls in this category and thus involves nothing truly mysterious. Just as a computational model of any real-world phenomenon does not imply that the model is actually that phenomenon (e.g., simulating a rainstorm does not make a computer wet [Searle, 1980]), simulating aspects of consciousness does not imply that the machine involved actually becomes conscious. In contrast, by *instantiated consciousness* we mean efforts to produce an artificial system that actually is phenomenally conscious, i.e., that experiences qualia and has subjective experiences and thus represents "synthetic phenomenology" [Chrisley, 2009]. Currently no existing work in AC has produced a generally accepted demonstration of instantiated machine consciousness, or even compelling evidence that instantiated machine consciousness is possible. Conversely there is currently no compelling theoretical or experimental proof that this will not be possible in the future. With this distinction between simulated and instantiated machine consciousness in hand, we can now return to the question about how work in AC may contribute to improving future AI systems.

A first answer to this question is that work on *simulated* consciousness is directly and immediately relevant to enhancing existing practical AI technology. For example, many existing AI systems are very brittle in the context of novel situations, including both AI systems based on traditional symbol processing methods and those based on contemporary deep learning methods (e.g., adversarial images for deep convolution networks). This is especially a problem with autonomous physical systems where a lack of trustworthiness, both in general but especially in the face of unanticipated novel situations, can be dangerous, and it has significantly limited the practical use of AI in such systems. There is substantial evidence that people, when confronted with novel situations, evoke

---

[e]Our simulated consciousness corresponds to MC1–MC3 and our instantiated consciousness to MC4 in the taxonomy of machine consciousness given in [Gamez, 2018].

conscious reasoning and learning to deal with these situations [Mudrik *et al.*, 2012]. This is true regardless of whether the situation is unexpected (e.g., a person driving a familiar highway route suddenly sees two cars collide up ahead) or simply a pre-planned novel experience (e.g., learning to ride a bike or play a game). Current AI systems also generally do not reflect on their internal models to reason about the causes of failure or difficulties. All of this suggests that AC studies relating consciousness to functions such as executive decision making, novelty detection, attention mechanisms, working memory, metacognition, motivations, and informing volitional activities appear promising avenues to explore in creating more effective ACI systems.

Another example of how AC studies of simulated consciousness may contribute to practical AI systems relates to the latter's interactions with people. Current human-computer interactions involving AI systems are quite limited. For example, there is no existing AI system that can consistently pass the Turing Test. Conscious self-monitoring would be expected to improve human-robot interactions because of the intimate relationship between self-awareness and the awareness of roles and perspectives [Trafton *et al.*, 2005]. In other words, understanding of roles in various situations is valuable in anticipating the behavior of others, and arguably this understanding relates to self-consciousness. Related ideas have been expressed concerning the importance of simulated inner speech to self-awareness and human trust of robots [Gerachi *et al.*, 2021]. These considerations suggest that simulated AC studies relating consciousness to functions such as working memory with its rapid one-step learning, self-awareness, self-modeling, source of motivations, and symbol grounding would be promising avenues to explore in developing ACI.

A second answer to the question about how work in AC may contribute to creating future AI systems relates to *instantiated* consciousness. While there is no generally accepted proof that instantiated machine conscious can or cannot be created, we speculate here about what it would mean for AI if a phenomenally conscious machine is someday possible. From a technological perspective, an ACI system based on instantiated consciousness would be anticipated to provide many of the same benefits of robustness, improved human-computer interactions, etc. as would an ACI system involving simulated consciousness. Further, Haikonen [2019] has compellingly argued that for an AI system to truly understand the outside world, its symbols must be grounded in qualia because qualia are self-explanatory forms of sensory information. In addition to these technological implications, perhaps even more significant would be how an instantiated machine consciousness would relate to the scientific study of consciousness. If we can successfully create and confirm an instantiated ACI, something that would effectively be the first true artificial mind, we will have made a fundamental advance in consciousness studies in general. Such an ACI would permit the study of consciousness at a much deeper level than is currently possible. For example, it would be expected to shed light not only on the core underlying mechanisms of consciousness, but also on improved criteria for rationally determining the presence/absence of consciousness in machines and animals, and the possibility of mind uploading. It might also lead to major advances in our understanding of psychiatric and neurocognitive disorders, such as schizophrenia, amnesia, and dementia.

7. Discussion

In this chapter we have asked the question of how concepts developed via computational models in AC might contribute to advancing the creation of more effective intelligent agents, or ACI, than can currently be supported by contemporary AI technology. We approached this question by reviewing past hypotheses about what the biological functions of consciousness are, most of which focus on the processing of information. In contrast, we hypothesized that short-term working memory and the associated very fast learning/unlearning of working memory's contents may also be considered as a function of consciousness, and one that complements/unifies previously suggested functions. In this context we reached three conclusions from our analysis.

First, work on simulated consciousness is immediately relevant to advancing the technology of AI, most prominently in terms of improving robustness and human-computer interactions. Particularly promising avenues for future ACI research in the short term can literally be read off from the list of previously hypothesized functions of consciousness in Sec. 2: detecting and managing novelty, symbol grounding and its impact on AI effectiveness, top-down executive control of behavior, the role of a self-model and motivations in machine intelligence, etc. Over the long term, ACI can be expected to play a major role in investigating the feasibility of speculative proposals about developing technologies such as mind uploading [Wiley, 2014] and digital ghosts [Steinhart, 2007].

Second, the development of instantiated machine consciousness (or MC4 [Gamez, 2018]), when combined with AI methodologies, would be a major technological and scientific advance that enables communicating with and studying in depth a conscious mind in ways that are currently not possible. Particularly exciting is the possibility of gaining insights into neurocognitive disorders involving anosognosia, such as aphasia, visual impairment following a stroke, spatial neglect, schizophrenia and dementia. The critical direction for future ACI research in this case is how to create an artifact that experiences qualia. At the current time there is no consensus on how this might be done, or even if it is possible.

Third, short-term working memory and especially the rapid learning/unlearning of its contents have been a largely overlooked possible function of consciousness. For the future, our own research is examining whether or not the three computational correlates of consciousness that have been suggested by AC models of working memory will be sufficient to support working memory *in increasingly general situations*. To examine this issue, we anticipate applying our methodology to challenging imitation learning tasks involving cause-effect reasoning where symbolic AI methods, but not neurocomputational methods, have previously been shown to work effectively [Katz *et al.*, 2018].

Of course, success in producing an ACI in any form would naturally raise several issues and concerns. The most immediate of these relates to the negative impact that increasingly successful ACI based on simulated consciousness could have on jobs and employment. There are already plenty of examples where existing AI technology has replaced human workers, such as the use of automated tellers and retail checkout stations. ACI could aggravate this widely discussed problem by allowing the use of AI technology to more broadly displace human workers (although this could be a positive factor when it shields humans from dangerous situations) or in situations where the need for better human-machine interactions currently limits their use, for example initial screening interviews of medical patients. More generally, a sufficiently powerful and general simulated ACI would raise fundamental scientific issues, such as how such an artificial automaton can even be distinguished from a truly phenomenally conscious biological entity (the *Blade Runner scenario*).

For the longer term, another prominent concern is what the development of an ACI based on instantiated consciousness would imply. Much attention has been given during recent years to the idea that ACI might lead us to a technological singularity that represents an existential threat to humanity (the *Terminator scenario*). It appears to us that a more plausible scenario is the physical integration of biological and machine intelligence, e.g., via brain-computer interfacing, something that is arguably starting to occur already and also carries great risks (the *Borg scenario*). A separate concern relates to what the rights of a phenomenally conscious machine would be, and this is also receiving attention. In some situations we would want to intentionally avoid creating a conscious machine, for example if we knew in advance that it would be treated badly or discarded. Further analysis and discussion of the tradeoffs between the benefits and risks of instantiated machine consciousness, and its ethical implications in general, is surely merited.

Acknowledgments

Our work on modeling working memory, one-step store-erase learning, the NVM, compositionality in working memory, and reinforcement learning was supported by ONR award N00014-19-1-2044.

References

Baars, B. [1997] In the Theatre of Consciousness, *Journal of Consciousness Studies*, 4, 292–309.

Baars, B., Franklin, S. [2003] How Conscious Experience and Working Memory Interact, *Trends in Cognitive Science*, 7, 166–172.

Baddeley, A. [2012] Working Memory, *Annual Review of Psychology*, 63, 1–29.

Bayne, T., Montague, M. (eds.) [2011] *Cognitive Phenomenology*, Oxford University Press.

Bringsjord, N. [2007] One Billion Dollars for a Conscious Robot, *J. Consc. Studies*, 14, 28–43.

Brody, J., Barham, S., Dai, Y., Maxey, C., Perlis, D., Sekora, D., Shamwel, J. [2016] Reasoning with Grounded Self-Symbols, *Artificial Intelligence for Human-Robot Interactions*, AAAI.

Carruthers, P. [2015] *The Centered Mind*, Oxford University Press.

Chella, A., Frixione, M., Gaglio, S. [2008] A Cognitive Architecture for Robot Self-Consciousness, *Artificial Intelligence in Medicine*, 44, 147–154.

Chrisley, R. [2009] Synthetic Phenomenology, *Intl. Journal of Machine Consciousness*, 1, 53–70.

Cleeremans, A. [2005] Computational Correlates of Consciousness, in S. Laureys (ed.), *Progress in Brain Research*, Vol. 150, pp. 81–98.

Collins, A., Frank, M. [2012] How Much of Reinforcement Learning is Working Memory, Not Reinforcement Learning? *European Journal of Neuroscience*, 35, 1024–1035.

Cowan, N., Elliott, E., Saults, J., Morey, C., Mattox, S., Hismjatullina, A., Conway, A. [2005] On the Capacity of Attention, *Cognitive Psychology*, 51, 42–100.

Davis, G., Katz, G., Gentili, R., Reggia, J. [2021] Compositional Memory in Attractor Neural Networks with One-Step Learning, *Neural Networks*, 138, 2021, 78–97.

DeLancey, C. [1996] Emotion and the Function of Consciousness, *J. Consciousness Studies*, 3, 492–499.

Devlin, J., Bunel, R., Singh, R., Hausknecht, M., Kohli, P. [2017] Neural Program Meta-Induction. In *Advances in Neural Information Processing Systems*, 2077–2085.

Earl, B. [2014] The Biological Function of Consciousness, *Frontiers in Psychology*, 5, 1–18.

Fodor, J., Pylyshyn, Z. [1988]. Connectionism and Cognitive Architecture: A Critical Analysis, *Cognition*, 28, 3–71.

Gamez, D. [2018] Human and Machine Consciousness, Open Book.

Geraci, A., D'Amico, A., Pipitone, A., Seidita, V., Chella, A. [2021] Automation Inner Speech as an Anthropomorphic Feature Affecting Human Trust, *Frontiers in Robotics & AI*, 8:620026. doi: 10.3389/frobt.2021.620026.

Graves, A., Wayne, G., Reynolds, M., Harley, T., Danihelka, I. *et al.* [2016] Hybrid Computing Using a Neural Network with Dynamic External Memory. *Nature*, 538 (7626), 471.

Haikonen, P. [2019] *Consciousness and Robot Sentience*, World Scientific.

Hupkes, D., Dankers, V., Mul, M., Bruni, E. [2020]. Compositionality Decomposed: How Do Neural Networks Generalize? *Journal of Artificial Intelligence Research*, 67, 757–795.

Holland, O. [2007] A Strongly Embodied Approach to Machine Consciousness, *J. of Consciousness Studies*, 14, 97–110.

Katz, B. [2013] An Embarrassment of Theories, *Journal of Consciousness Studies*, 20, 43–69.

Katz, G., Davis, G., Gentili, R., Reggia, J. [2019] A Programmable Neural Virtual Machine Based on a Fast Store-Erase Learning Rule, *Neural Networks*, 119, 10–30.

Katz, G., Gupta, K., Reggia, J. [2020] Reinforcement-Based Program Induction in a Neural Virtual Machine, *Proc. 2020 International Joint Conference on Neural Networks (IJCNN)*, 1–8, doi: 10.1109/IJCNN48605.2020.9207671.

Katz, G., Huang, D., Hauge, T., Gentili, R., Reggia, J. [2018] A Novel Parsimonious Cause-Effect Reasoning Algorithm for Robot Imitation and Plan Recognition, *IEEE Transactions on Cognitive and Developmental Systems*, 10, 177–193.

Kruijne, W., Bohte, S., Roelsema, P., Olivers, C. [2021] Flexible Working Memory Through Selective Gating and Attentional Tagging, *Neural Computation*, 33, 1–40.

Kuipers, B. [2008] Drinking From the Firehose of Experience. *Artif. Intell. Medicine*, 44, 155–170.

Lake, B., Baroni, M. [2018]. Generalization Without Systematicity: On the Compositional Skills of Sequence-to-Sequence Recurrent Networks, *International Conference on Machine Learning*, 2873–2882.

Lake, B., Ullman, T., Tenenbaum, J., Gershman, S. [2017]. Building Machines that Learn and Think Like People, *Behavioral and Brain Sciences* 40.

Lara, A., Wallis, J. [2015] The Role of Prefrontal Cortex in Working Memory, *Frontiers in Systems Neuroscience*, December.

Lin, R., Katz, G., Reggia, J. Effectiveness of Proximal Policy Optimization Methods in Training a Neural Virtual Machine, *Proc. 23rd International Conference on Artificial Intelligence* (ICAI'21), July 26–29, 2021, in press.

Manohar, S., Zokaei, N., Fallon, S., Vogels, T., Husain, M. [2019]. Neural Mechanisms of Attending to Items in Working Memory, *Neuroscience and Biobehavioral Reviews*.

Manzotti, R. [2012] The Computational Stance is Unfit for Consciousness, *International Journal of Machine Consciousness*, 4, 401–420.

McDermott, D. [2007] Artificial Intelligence and Consciousness. In *Cambridge Handbook of Consciousness*, M. Moscovitch and E. Thompson (eds.), Cambridge University Press, 117–150.

Metzinger, T. [2000] Neural Correlates of Consciousness, MIT Press.

Mudrik, L., Deouell, L., Lamy, D. [2012] Novelty, Not Integration: Finding the Function of Conscious Awareness, in S. Kreitler and O. Maimon (eds.), *Consciousness: Its Nature and Functions*, Nova Science, pp. 265–276.

Nefdt, R. [2020]. A Puzzle Concerning Compositionality in Machines, *Minds and Machines*, 1–29.

Pasquali, A., Timmermans, B., Cleeremans, A. [2010] Know Thyself: Meta-Cognitive Networks and Measures of Consciousness, *Cognition*, 117, 182–190.

Perlis, D. [1997] Consciousness as Self-Function, *Journal of Consciousness Studies*, 4, 509–525.

Persuh, M., LaRock, E., Berger, J. [2018] Working Memory and Consciousness, *Frontiers in Human Neuroscience*, 12, March.

Pierson, L., Trout, M. [2017] What is Consciousness For?, *New Ideas in Psychology*, 47, 62–17.

Pratte, M. [2018] Iconic Memories Die a Sudden Death, *Psychological Science*, 29, 877–887.

Reggia, J. [2013] The Rise of Machine Consciousness, *Neural Networks*, 44, 112–131.

Reggia, J., Huang, D., Katz, G. [2017] Exploring the Computational Explanatory Gap, *Philosophies*, 2, 5, doi:10.3390/philosophies201005.

Reggia, J., Katz, G., Davis, G. [2018] Humanoid Cognitive Robots that Learn by Imitation, *Frontiers in Robotics and AI*, Section 5.

Reggia, J., Katz, G., Davis, G., Gentili, R. [2021] Avoiding Catastrophic Memory Loss with Short-Term Working Memory, *Proc. 23rd International Conference on Artificial Intelligence*.

Reggia, J., Monner, D., Sylvester, J. [2014] The Computational Explanatory Gap, *Journal of Consciousness Studies*, 21, 153–178.

Rose, N., LaRocque, J., Riggall, A., Gosseries, O., Starrett, M., Meyering, E., Postle, B. [2016]. Reactivation of Latent Working Memories with Transcranial Magnetic Stimulation, *Science*, 354, 6316, 1136–1139.

Rosenthal, D. [2008] Consciousness and its Function, *Neuropsychologia*, 46, 829–840.

Russell, S., Norvig, P. [2010], Intelligent Agents, Chapter 2 in *Artificial Intelligence: A Modern Approach*, Prentice Hall, 34–63.

Sanz, R., Hernandez, C., Sanchez-Escribano, M. [2012] Consciousness, Action Selection, Meaning and Phenomenic Anticipation, *International Journal of Machine Consciousness*, 4, 383–393.

Searle, J. [1980] Minds, Brains, and Programs. *Behavioral and Brain Sciences*, 3, 417–424.

Seth, A. [2009] The Strength of Weak Artificial Consciousness, *Intl. J. Machine Consiousness*, 1, 1–82.

Shanon, B. [1998] What is the Function of Consciousness, *J. Consc. Studies*, 5, 295–308.

Squire, L., Dede, A. [2015] Conscious and Unconscious Memory Systems, *Cold Spring Harbor Perspectives in Biology*, 7.

Steinhart, E. [2007] Survival as a Digital Ghost, *Minds & Machines*, 17, 261–271.

Sun, R., Franklin, S. [2007] Computational Models of Consciousness. In Zelazo, P., Moscovitch, M. (eds.), *Cambridge Handbook of Consciousness*, Cambridge University Press, 151–174.

Sylvester, J., Reggia, J. [2016] Engineering Neural Systems for High-Level Problem Solving, *Neural Networks*, 79, 37–52.

Sylvester, J., Reggia, J., Weems, S., Bunting, M. [2013] Controlling Working Memory with Learned Instructions, *Neural Networks*, 41, 23–38.

Takeno, J. [2013] *Creation of a Conscious Robot*, Pan Stanford.

Taylor, J. [2007] CODAM: A Neural Network Model of Consciousness, *Neural Netw.*, 20, 983–92.

Tononi, G. [2008] Consciousness as Integrated Information, *Biological Bulletin*, 215, 216–242.

Trafton, J., Cassimatis, N., Bugajska, M., Brock, D., Mintz, F., Schultz, A. [2005] Enabling Effective Human-Robot Interaction Using Perspective-Taking in Robots, *IEEE Trans. Systems, Man & Cybernetics*, Part A, 35, 460–470.

Turing, A. [1950] Computing Machinery and Intelligence. *Mind*, 59, 433–460.

Wiley, K. [2014] A Taxonomy and Metaphysics of Mind-Uploading, Alauton Press.

Williams, R. J. [1992] Simple statistical gradient-following algorithms for connectionist reinforcement learning, *Machine learning*, 8(3), 229–256.

Chapter 9

Will Conscious Digital Creatures Roam the Metaverse?

Owen Holland

Sussex Centre for Consciousness Science, University of Sussex,
Brighton BN1 9RH, United Kingdom
o.e.holland@sussex.ac.uk

This paper is an extended version of "Forget the Bat" (Holland, 2020). The philosophical legacy of the idea that there must be something it is like to be a conscious organism, together with an inclination to regard organic life forms as having qualities fundamentally distinct from other physical systems, have adversely affected the development of the nascent discipline of machine consciousness. The paper highlights some of the factors involved, and as a corrective proposes a reframing of machine consciousness within a thoroughgoing engineering context. This is shown to offer some significant avenues for progress, while bringing with it some theoretical problems requiring further consideration. In particular, recent work by others on the putative consciousness of wholly virtual systems points to the possibility of using the developing metaverse to host and enable interaction with conscious digital creatures.

Keywords: metaverse; artificial consciousness; virtual reality

1. Introduction: The Metaverse

"No army can resist the force of an idea whose time has come," wrote Victor Hugo. It is difficult to disagree with this observation, but unfortunately its validity in respect of any given idea can only be established after its force has clearly not been resisted. What it is not is a guide to spotting the idea whose time will come early enough to prepare for it in a practical way, such as investing in it, or planning to exploit it in a different context than the one that inspired its development.

If the much-discussed Metaverse is such an idea, exactly what will it consist of? This is not yet clear — some claim it already exists, and some that it could never exist, but the future Metaverse is generally thought of as an alternative entirely virtual world or universe immersively accessible for business or pleasure to technologically equipped humans via their avatars (virtual representations of themselves). It will be loosely modeled on the real world and will have links to and be the subject of real-world laws and money systems. It should be persistent (once started it should never stop), synchronous (using the same time as real-time), and may even consist of a number of different but interoperable Metaverses. Its credibility as an idea is strongly rooted in the success of the internet — indeed, some of its detractors dismiss it as "a 3-D internet".

The name and many of the associated concepts of the Metaverse are now 30 years old — the original source is the 1992 science fiction novel "Snow Crash", written by Neal Stephenson [Stephenson, 1992]. Is 2022 close to the year when its time may come? If it is not, there may be some very heavy investment losses in the meantime — Meta (formerly Facebook) is currently spending at the rate of $10 billion a year on the enabling technologies for its Metaverse and is looking at a 5 to 10-year interval before the potential rewards (to Meta) are seen. However, there were only 30 years between 1948, the date of demonstration of the first program in history (all seventeen instructions of it) to run on a digital, electronic, stored-program computer (the Manchester "Baby"), and 1978 when the launches of the Apple II and the DEC VAX can be seen as marking the release of the irresistible force of the idea of digital computation in homes and businesses. 34 years took us to the launch of the IBM PC, which commoditized the idea almost into invisibility, and after 41 years we saw the essential completion of the hardware and software infrastructure of the internet and the World Wide Web, with its novel and explosive financial, social, and political consequences. The age of the Metaverse may at least be close enough for us to think about what it could be used for in contexts other than those envisaged by its early proponents and component manufacturers.

In "Snow Crash" the fictional Metaverse was predicated on the use of avatars to enable people to access the vicarious first-person experience of a virtual world in preference to the real experience of a dystopic world and culture. Among many other things, the novel described how the same unbridled commercial forces that had corrupted the real world had worked to corrupt the virtual world into an alternative and presumably less dystopic experience. The context of interest to this paper is not one of politics, commerce, crime, or addiction to virtual reality experiences, but that of artificial consciousness: could the virtual world of the Metaverse include artificially conscious entities enabling first-person human interactions with them, just as the real world includes naturally conscious entities in the form of humans and almost certainly other species?

In Stephenson's Metaverse, human-controlled avatars interacted with other human-controlled avatars and with objects in the virtual environment, but other than the avatars there were no other active intelligent agents, and certainly none that were or appeared conscious. A curiosity is that Stephenson acknowledged one of the main inspirations for the book as being Julian Jaynes' "The origin of consciousness in the breakdown of the bicameral mind" [Barnes and Noble, 2008]. Some idea of how "Snow Crash" informed and influenced technology development can be gained from the fact that some leading companies made "Snow Crash" recommended or compulsory reading for their research staff. Its influence on some early virtual world software, such as Second Life, has been acknowledged by their creators [Maney, 2007]. It penetrated even into academia — an extract from a 2002 textbook dealing with the virtual objects expected to be present in a given virtual environment reads as follows:

"A special subset of the objects is called actors. An actor is an object that can initiate an interaction with another object or actor…There is a special actor, a human operator, or participant in the environment. The participant has a visual representation within the environment (e.g., a cursor in the simplest possible graphical application, a full humanoid representation, or an avatar, in more sophisticated applications). A more complex example was provided by Neal Stephenson in his…novel Snow Crash. The Metaverse is a three-dimensional virtual world that humans can enter into embodied as "Avatars" — typically 3D graphical representations of the human body or something approximating this. But some objects, supposedly Avatars, were really system Daemons representing processes carried out by the computer systems maintaining the Metaverse in human form" [Slater *et al.*, 2002].

The absence of higher level autonomous virtual agents in "Snow Crash" does not mean that such agents were not the concern of the contemporary technology industry. Known as Non-Player Characters (NPCs) from their origins in role-playing computer games, they are programmed to exhibit certain behaviors in response to certain situations. While they are usually behaviorally simple, there need be no limit to their complexity, and they may be programmed to appear intelligent and even human-like to the human player. Intriguingly, the NPC rated most humanlike at a gaming competition in 2009 [Arrabales, 2020] had a design based on Baars' Global Workspace Theory of consciousness [Baars, 1988]; however, conventional AI and the relatively new methodology of behavior trees form the main framework of today's complex NPCs, with the focus being on the impression that is created by the agent's behavior rather than on mimicking any mechanisms thought to be used by intelligent conscious humans. (In fact, NPC is used in some circles as an insult to someone who appears incapable of thinking for himself.) NPCs are perfectly compatible with Stephenson's Metaverse, as the system Daemon avatar makes clear, but are simply not featured.

The main question in our paper is this: Will the digital substrate of the Metaverse be capable of supporting autonomous conscious entities along with the avatars of conscious humans?

2. The Reality of Virtual Reality

We are fortunate that David Chalmers, in his recent book "Reality+: Virtual Worlds and the Problems of Philosophy" [Chalmers, 2022] has dealt with some of the issues surrounding our question. He declares in his Introduction that, "The central thesis of this book is: Virtual reality is genuine reality." This challenging position leads to a thorough analysis from a philosophical point of view of the extent to which virtual reality is or is not an illusion, the factors bearing on how real virtual reality feels, the role of the base technology of computer simulation, the nature and presence of Non-Player Characters, and many other perspectives. In addition, he confronts part of our question directly, in Chapter 15: "Can there be consciousness in a digital world?"

As a leading philosopher of consciousness, he is initially surprisingly frank about the general position: "Why is there consciousness in the universe? How do physical processes give rise to consciousness? How can there be subjective experience in an objective world? Right now, no one knows the answers to these questions." He then narrows his question down to "It's hard to know for sure whether silicon machines can be conscious. One reason is that it's hard to know for sure whether any entity other than oneself is conscious." This brings in the classic philosophical problem of other minds. He then declares that "The problem of machine consciousness is an especially hard version of the problem of other minds" and proceeds to finesse the issue of other minds with a further narrowing: "I'll concentrate on just one type of machine: a perfect simulation of a brain, such as my own. The brain simulation is a digital system running on a computer. If we can establish that one digital system is conscious, then we know there's no general reason why digital systems cannot be conscious, and the floodgates will open."

He goes on to claim that the most plausible hypothesis about the outcome of a gradual mind uploading thought experiment using such a simulation coupled to the original brain's body is "…that simulated brains can be conscious. At least in the special case in which you become a simulated brain by gradual uploading, the simulation will be fully conscious." While his arguments meet his own criteria for proposing and claiming the reality of simulated systems and worlds (he also writes at length about simulations in a rather broader context) and philosophically validate the idea of the reality of simulated minds, by design the book stops short of also being a technically adequate account of the possible existence of artificially conscious agents in a future Metaverse environment. The rest of this paper is intended to provide an engineering-based view of the digital consciousness problem, which when combined with Chalmers' philosophical approach may enable a fruitful and positive triangulation of the relevant issues.

3.   Forget the Bat

It is a sad fact that the emerging discipline known as machine consciousness, or artificial consciousness, does not seem to be making any progress. At least partly because the level of recent activity is so low, Reggia's [2013] review is still largely current. This is in stark contrast to the neuroscience of consciousness, where new papers, experiments, and findings form a gusher of information and at least partial understanding. However, both have something important in common: neither has yet delivered an adequate explanation of exactly what phenomenal consciousness is, or how it arises. Manzotti and Chella [2018] claim that what they call 'Good Old-Fashioned Artificial Consciousness' (GOFAC) has failed because it "suggests a physical world in which consciousness appears as a result of a specific intermediate level. A theory based on the idea that consciousness emerges from an intermediate level should explain what this level is and why it produces consciousness....This approach can be named the 'intermediate level fallacy' and it seems to be attractive because the introduced level appears less intimidating and more familiar than consciousness itself." In fact, the same can be said of many of the neuroscientific theories of consciousness — they identify the presence or activity of specific structures or processes as constituting consciousness itself, rather than being mere correlates or causal factors of phenomenal consciousness. Of course, illusionism [Frankish, 2017] and purely functional cognitive theories of consciousness such as Global Workspace Theory [Baars, 1987] do not need to commit the fallacy because for them there is nothing real requiring explanation.

It goes without saying that the development of machine consciousness would be made significantly easier if a satisfactory theory of phenomenal consciousness existed and was specified in enough detail that engineers could simply press on with its implementation. In the context of a cognitive theory of consciousness, the late Stan Franklin's implementations of Baars' Global Workspace Theory [Franklin, 2003] show exactly what sort of process and level of detail are needed. For neural theories of consciousness, Edelman's series of robots [2007] provided significant support for at least some elements of his scheme. Unfortunately, most theorists of consciousness do not seem to appreciate just how concrete the expression of a theory must be for implementation even to be feasible, and equally unfortunately very few engineers have a good enough knowledge of consciousness studies to produce an implementable theory that could deal adequately with the intrinsic challenges. In order to chart a credible future course, it may be helpful to go back to the beginning of what we could call the modern period and examine the explanandum — phenomenal consciousness itself — with the eye and mind not of a philosopher or neuroscientist or biologist, but of an engineer.

4. The Philosophical Bat Problem

We can date the modern period by the sudden rise in articles pointing to phenomenal consciousness using the verbal formula "There is something it is like to be (something)". Although Farrell [2016] identifies several earlier uses of the phrase, it was the philosopher Thomas Nagel's seminal paper "What is it like to be a bat?" [1974] that fueled its almost compulsory use in recent discussions. Nagel wrote:

"Conscious experience is a widespread phenomenon. It occurs at many levels of animal life, though we cannot be sure of its presence in the simpler organisms, and it is very difficult to say in general what provides evidence of it. (Some extremists have been prepared to deny it even of mammals other than man.) No doubt it occurs in countless forms totally unimaginable to us, on other planets in other solar systems throughout the universe. But no matter how the form may vary, the fact that an organism has conscious experience at all means, basically, that there is something it is like to be that organism. There may be further implications about the form of the experience; there may even (though I doubt it) be implications about the behavior of the organism. But fundamentally an organism has conscious mental states if and only if there is something that it is like to be that organism — something it is like for the organism."

Farrell [2016] dubs this and the discussions it inspired 'WIL-talk' ('What it is Like' talk) and after analyzing it from a philosophical perspective arrives at the conclusion that "There is, as yet, no consensus about how to understand WIL-talk". My immediate concern as an engineer is not with the What-It-Is-Like part of the concept, but with the other half of the usual phrase — in the case of Nagel's article, "to be a bat".

Suppose I have a perfectly good but unconscious robot, and I then develop a consciousness module and fit it to the robot, connecting it appropriately to sensor and motor circuits, batteries, and so on. What sort of sense would it make for me to ask if there is now something it is like to be that robot? There will certainly be a something-it-is-like process going on in the consciousness module, necessarily involving communications from and to the unconscious robot, but the module is just a physical and functional part of the robot, not the whole thing. To say that there's now something that it is like to be that robot would be to fall into the fallacy of composition — the opposite of the mereological fallacy — because the properties of the part are being attributed to the whole. Now if instead of wiring the consciousness module directly to the robot, I put it on the bench and make the necessary connections by wireless, there will still be the same something-it-is-like process going on, but the module will not be a physical part of the robot, although it will still be a functional part. If I then migrate the internal digital circuitry of the module to the cloud, the same something-it-is-like process will still be going on, but there will be no physical consciousness module as such for it to be going on in, just some bits of software distributed over several remote computers and being moved around depending on the availability of memory and processor cycles. However, the something-it-is-like process and the what-it-is-like will be identical in all three of these cases, conditioned entirely by the robot's sensors, motors, and environment. Arguably, the process will seem to itself to be the robot in all three cases in exactly the same sense that the bat's consciousness is busy seeming to itself to be the bat, but to say that there is something it is like to be the robot will be just as illegitimate as saying there is something it is like to be the bat, unless the verbal construction is understood as meaning something other than what it says. (Farrell [2016] discusses and largely dismisses the claim that WIL-talk is used in a technical sense.) But it might be perfectly correct to say that somewhere in the bat — most likely just in the brain, though some would disagree — there is something it is like to seem to be a bat.

5. The Biological Bat Problem

Biologists often make the point that consciousness is a biological phenomenon, and needs to be treated as such, i.e., treated differently from phenomena in other physical systems. While certainly legitimate within a biological context (see for example Ginsburg and Jablonka [2019]), this kind of statement usually cuts no ice with engineers, especially those working in the area of biologically inspired robotics who may have been successful in building and exploiting engineered analogues of biological structures. Nevertheless, one aspect of phenomenal consciousness is that at present it is only found in (certain) living creatures, and so must have been produced by and therefore may play some role in evolution. If phenomenal consciousness as such has some function that has contributed to evolutionary success, then it would certainly be of interest to engineers to know what that function is. There are several distinct possibilities bearing on this. The phenomenal aspect could be an epiphenomenal consequence of the particular biological implementation of the functional cognitive aspect that plays a net positive role in evolution. For example, as Humphrey [2011] and others have speculated, phenomenal consciousness may provide some new intrinsic motivations towards survival and self preservation over and above the particular instinctive and reflex mechanisms of the species that keep individuals alive long enough for successful reproduction. It has to be said that the idea of the prolongation of individual life as being an unalloyed good is incompatible with the operation of evolution, which is based on successful reproduction in a competitive and hostile environment where, to put it crudely, passengers are not carried. To some extent this belief seems to be a consequence of the popularity of the idea of autopoiesis, some advocates of which may not adequately recognize the primary roles played in evolution by reproduction and selective death.

Another quasi-biological perspective is an emphasis on the unique nature of living beings, and the properties related to cognition and consciousness that might arise directly from this nature. For example, Evan Thompson writes: "Life and mind share a common pattern of organization, and the organizational properties characteristic of mind are an enriched version of those fundamental to life. Mind is life-like. But a

simpler and more provocative formulation is this one: Living is cognition."
[Thompson, 2009] While an engineer can understand the rather different
perspective from which Thompson is writing, it is difficult to extract a
concrete meaning from his work that would be useful in designing an
artificial cognitive or conscious system.

Living creatures are also evolved wholes, and the wholeness is
sometimes seen by biologists and others as an intrinsic characteristic,
giving rise to a reluctance or refusal to divide them into functional parts,
such as brain and body. For example, witness the recent rise of the ideas
of embodied cognition, morphological computation, and the enbrained
body. However, for the engineer, who practices synthesis rather than
analysis, every apparent engineering whole, such as an airliner, is just a
collection of parts, and while they will be designed to work together
harmoniously, the only relevance of a concept of wholeness is in their
maintaining adequate functionality and in their sharing a common fate.

In living conscious creatures, the structures and processes supporting
consciousness are currently believed to be neurons. Engineers will not be
using biological neurons in attempting to create artificial consciousness
but will mainly use synchronous digital systems — conventional
computers — perhaps with some asynchronous and analogue subsystems
for special functions. They may use the digital systems to simulate
assemblies of neurons to any desired degree of accuracy, as suggested by
Chalmers [2022], or they may use any other approach they judge to be
appropriate for the task. In the longer term quantum computing may be
used, but we will not speculate about any additional properties this might
offer other than more speed. It is certainly possible that some presently
unknown properties of assemblies of neurons in biological bodies may be
responsible for phenomenal consciousness, and also for some presently
unknown cognitive aspects of consciousness, and that these properties
may not be reproducible using conventional digital techniques.

All in all, bats are historically and disciplinarily a troublesome starting
point for a technology of machine consciousness, so we should perhaps
begin with the engineering possibilities and constraints, and work
backwards towards whatever natural phenomenal consciousness may be.
This brief paper can contain no more than a sketch of how to proceed, but
should at least provide some grounds for realistic optimism.

6.   There is Something That it Seems Like

We are almost done with the bat, but not quite. A point where an engineer can constructively differ with Nagel comes from the sentence in the quotation above: "There may be further implications about the form of the experience." It is not completely clear what he means by this, but what he does not mean is clarified in his footnote to the sentence: "...the analogical form of the English expression "what it is like" is misleading. It does not mean "what (in our experience) it resembles," but rather "how it is for the subject himself."" However, once we leave the bat behind, we find that the situation with a putatively conscious digital machine is rather more favorable, in that we can know everything about its internal processes and their influence on the subsequent development of further internal processes and any actions resulting from those processes. (We note here that if the digital machine takes input from and sends output to an exclusively digital environment, this also can be known completely and exactly, unlike the real physical world.) There can be no privacy, and no doubts about the system's functional characteristics. More particularly, there is nowhere for subjectivity to hide, because it is clearly arguable that any subjective view must be represented, and anything that is represented can in principle be known — in other words, we might be able to say "what...it resembles" by giving a more or less exact and complete description. This would correspond to "what it seems like" — something not important for Nagel within the context he sets out, but important for us, as subjective experience is always about seeming rather than about being, even if the seeming corresponds accurately to some objective situation in the system's environment. It would certainly be inadmissible to claim that privacy must be an essential quality of subjectivity in all possible conscious systems, although it is certainly a contingent quality in a complex organism such as a bat.

But if something is somehow experienced by a digital system, must it be represented? At present a reasonable answer is "Yes, in some form" while being strictly non-committal about the form of the representation, which may require work to make it intelligible to us. In "Why and How Does Consciousness Seem the Way it Seems?" When Dennett [2015] discusses what he sees as problems with the idea of qualia, he stresses that in his view perceptual processes do not need to undergo some transduction or rendering to produce alleged qualia — the currency of experience — but rather function directly at the neural level of representation. However, in attempting to represent the subjective experience of a digital system to ourselves, we may well need to subject the record of the experience to some transduction or rendering in order to describe it.

A useful illustrative example comes from the CRONOS project [Holland *et al.*, 2007; Marques and Holland, 2009], a robot-based investigation of a possible machine consciousness implementation. The anthropomimetic robot CRONOS was equipped with a designed self model and world model capable of predicting the consequences of interactions between the robot and the world by using physics-based simulation. While no claims were made about any phenomenal consciousness within the system, the locus of interest was in the current state of the self and world models, and in the transactions between them. The self model and world model were first aligned with the current self and world physical states, and then offline simulations searched for actions that would fulfil the currently dominant motivation. It was possible to view the simulations (and the controlling code) so that the internal processes were completely known and could be understood. Moreover, the evaluation of the outcome of a planned action was done by re-applying to the simulated final state the visual perception software dealing with the initial capture of the real-world state, in line with Hesslow's [2002] identification of conscious thought as involving the re-use of perceptual and behavioral mechanisms. In Dennett's terms, this further processing of a rendered view of an internal process could be seen either as analogous to the use of qualia, or as a demonstration that all the necessary work was being done by code (rather than neural circuitry) and no intermediate process of transduction into a conscious representation was necessary.

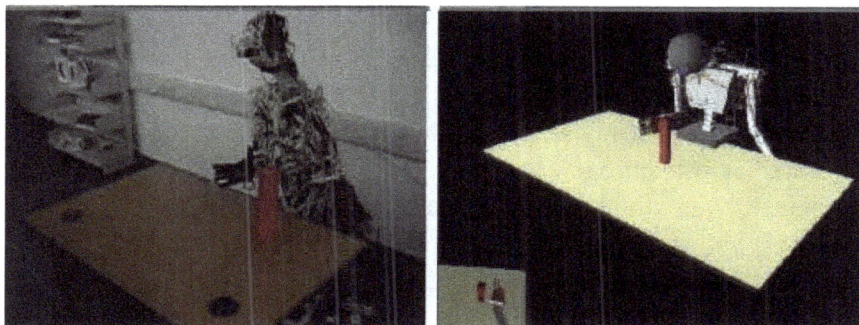

Fig. 1. On the left is the robot CRCNOS in its deliberately simple environment. On the right is the third-person rendered view of the physics-based simulation of CRONOS in its environment. The small window in the bottom left-hand corner is the rendered view from the position of the robot's eye — if the robot were conscious, this might correspond to the contents of consciousness, but all the information in the third person view would be available to the system.

Of course, in conscious humans and in future conscious machines, not all representations need to be or will be experienced, and it is the nature and mechanism of the selection of representations to be experienced that is one of the keys to understanding consciousness. If consciousness has a function, then representations that are experienced must have different consequences than those not experienced, and many of these differences are known, such as their influence on long term memory, or their ability to enable trace conditioning. In the CRONOS system, the formation of the inner representation from the incoming sensory information is made clear, as is the way in which the inner representation can be exploited by being manipulated at its own level and used to inform a decision about action. Although simple and crude, this scenario illustrates the operation of some of the components that must be in place before judgments can be made about which parts of the process may serve as candidate experiences and which parts may not.

The fundamental importance to consciousness of seeming rather than being is currently being realized in many ways. The most extreme manifestation is in illusionism [Frankish, 2017] where the subjective reality of consciousness is understood as mere seeming. The sort of argument this can give rise to between a Realist and an Illusionist is illustrated by Chalmers [2018] as follows:

"Illusionist: My view predicts that you will find my view unbelievable, so your denial simply confirms my view rather than opposing it.
Realist: I agree that my denial is not evidence against your view. The evidence against your view is that people feel pain.
Illusionist: I don't think that is genuine evidence.
Realist: If you were right, being me would be nothing like this. But it is something like this.
Illusionist: No. If 'this' is how being you seems to be, then in fact being you is nothing like this. If 'this' is how being you actually is, then being you is just like this, but it is unlike how being you seems to be."

While we may not be able to argue with an artificial conscious system in the foreseeable future, we should be able to access and perhaps comprehend how things may seem to a prototype artificially conscious system — how in a sense it represents them to itself — and to compare them to whatever objective reality we have imposed on the system.

7.   What Would a Theory of Consciousness Be Like?

If we had a satisfactory general theory of phenomenal consciousness, what general form could it take? One simple possibility is that it could be like this: In a system of type X, structures or processes Y have properties or capacities Z, where Z describes the properties of phenomenal consciousness. As far as human consciousness is concerned, type X could be a normal awake adult human, Y would be the structures and processes involved in the production of phenomenal consciousness, and Z would be the description of the properties and capacities of Y in terms of human phenomenal consciousness. At present, of course, the biggest problem is the articulation of Z, as we have no agreed adequate description. However, since many theories of consciousness — for example, Global Workspace Theory [Baars, 1987] and its subsequent variants — are not in themselves theories of phenomenal consciousness, but of the cognitive correlates of consciousness, they could be described using a Z restricted to non-phenomenal aspects. Some theories may use awareness, self awareness, and minimal consciousness as weasel words when referring to purely cognitive characteristics, and some may justify these usages by asserting that the cognitive aspects are constitutive of consciousness, or that phenomenal consciousness is a cognitive illusion and so the cognitive correlates of the illusion are therefore sufficient, but all of these would again only require a non-phenomenal Z. In some theories, such as Information Integration Theory 3.0 [Oizumi *et al.*, 2014], a type X consisting of digital systems would preclude any possibility of computer based phenomenal consciousness, and so again a non-phenomenal Z would have to be used.

The above formulation has the virtue of accommodating a strategy of approaching artificial consciousness by degrees by first designing and testing systems that attempt to meet the cognitive requirements of phenomenal consciousness using a non-phenomenal Z, thus focusing attention on what, if anything, is still missing. It is certainly possible that, in some yet to be understood way, cognitive characteristics may combine to form a system supporting phenomenal consciousness, or that this factor combined with some particular implementation forms such a system. Both of these legitimize an approach of building a system that implements the cognitive correlates, and then seeing whether it has any unforeseen emergent properties pointing the way to phenomenal consciousness. Although Franklin's [2003] IDA system was not designed to be part of this strategy, it does offer an example of the skills and knowledge called for in a satisfactory implementation, which go beyond philosophical, neuro-scientific, and psychological disciplines, and Franklin does carefully discuss the scope of the project in relation to phenomenal consciousness.

The reference to above to normal adult humans as representing the current type X for human phenomenal consciousness implicitly contains a normative element, as it allows outliers to be discarded. However, engineers are very wary of discounting the behavior of aberrant or malfunctioning systems because they may point to unrecognized systemic or intermittent flaws in a designed system, thus providing a broader context for the understanding of a system's characteristics. If you asked each of 100 competent electronics engineers to design a high-fidelity amplifier to meet a particular specification, and then tested the systems to see if they actually met the specification, all would do so. However, if you damaged the systems in various ways, or provided out of specification inputs (signals too strong, too weak, including frequencies beyond the limits of the specification, or an environment too hot or too cold or with too much electronic noise, and so on) the systems would fail in various ways, but the manner of their failure would depend on the particular design choices each engineer had made, and so would give information about those choices. In essentially the same way, the characteristics of consciousness in human systems that are genetically, biochemically, medically, or developmentally compromised, or physically damaged by accident or surgery, constitute a source of information not just about the consciousness of normal humans but about the underpinnings of consciousness in general. Perhaps one of the most important current issues in the neuroscience of consciousness is the status of Merker's [2007] contention evidenced by anencephalic and hydranencephalic patients that primary consciousness is produced by subcortical structures, and that thalamocortical loops add content and complexity. An implementation of Merker's model would be much simpler and more tractable that tackling full-blown normal consciousness, and possible represents the best way forward at present.

8.   A practical engineering perspective on the reality/virtuality debate

Although machine consciousness offers opportunities denied to theoretical analyses of consciousness derived from biological examples, it is also subject to its own constraints, many involving practical rather than theoretical issues — at least on the surface. As Holland wrote [2019]: "Whether we are to create an artificial consciousness in order to understand natural consciousness, or to provide a basis for a technology of consciousness, we have a choice of doing it either in a real robot in the real world, or in a virtual robot in a virtual world (or if experiencing social interaction with conspecifics is necessary to become conscious, many real or virtual robots), or in both. From the point of view of efficiency, tractability, access to information, and sheer practicality, the virtual option beats the real option hands down. Working with real robots and the real world is slow and difficult: robots break, age, wear out, need updating, are expensive, and need constant engineering attention; the real world is incompletely knowable, and also changes with time. Further, it is impossible to speed up real time, to which robots and the real world are unavoidably confined, or to repeat an experiment exactly. With a virtual approach, changes are easy, knowledge of the robot and the world is in principle complete, multiple parallel implementations can be run, real time is not a constraint, and the base technology improves at an exponential rate." Details of possible methodologies for extending a virtual conscious system into a physical artefact in the real world can also be found in [Holland, 2019].

A further possible advantage of a completely virtual approach is that, in the absence of a comprehensive design strategy for a conscious artefact, it potentially allows the use of artificial evolution to investigate the appearance of analogues of features of consciousness in modelled entities subjected to analogues of the pressures of natural evolution.

These overwhelmingly practical observations raise the issue of the physicality or virtuality of the system providing the sensory input and effector output components for the conscious agent, where the interesting question is whether a completely virtual simulated system in a simulated environment could support real phenomenal consciousness, by for example feeling real pain? This is of course part of what Chalmers [2022] set out to clarify from a philosophical viewpoint as summarized in Section 2 earlier. (Note that Chalmers' realist above [Chalmers, 2018] chose the ability to feel pain as the touchstone for the reality of consciousness.) While this example points to the central problem very clearly, we should reflect on the fact that congenital insensitivity to pain is a very rare but well-known condition in which the sensitivity to potentially injurious or painful stimuli is normal, but the patient does not feel pain. Various genetic causes have been found in some cases (e.g. [Habib *et al.*, 2019]) but the patients' consciousness is otherwise normal, so the capacity for feeling pain in itself is not a necessary requirement for phenomenal consciousness, although the opposite is of course true. However, anecdotal evidence points to the emotional tone of some patients being atypically positive, probably associated with an excess of endorphins and other psychoactive substances. Hypnotic analgesia and placebo analgesia are also well-established phenomena, and brain scans during instances of such analgesia show changes within the brain similar to those occurring during chemical analgesia, so degrees of genuine insensitivity to specific pains can apparently be produced in normal humans. It is not yet clear whether there is any satisfactory choice for a fundamental characteristic of consciousness that would not fall victim to some version of this counter-example.

In relation to this physicality/virtuality problem, discussions with a variety of consciousness researchers reveal a frequently expressed viewpoint of what is still a hypothetical but unduly neglected situation: that the same computational system that would produce real consciousness in a physical robot would not produce real consciousness in an equivalent virtual robot — some element of physicality seems to be seen by many as an intrinsic and necessary constituent of consciousness. While reasoned arguments to support this position are elusive, Chalmers' arguments in favor of the reality of simulated minds may be effective in opposing it. He summarizes his conclusions in Chapter 15 as follows: "All this strengthens the case that virtual reality is genuine reality, by making the case that simulated minds are genuine minds. These arguments also support the prospects for artificial consciousness more generally, whether or not the artificial system involves a simulated brain" [Chalmers, 2022].

9.   Conclusion

We can assume that the Metaverse, when it appears in some adequately developed form, will be able to host digital Non-Player Characters of arbitrary complexity that may have any desired avatar-like appearance and characteristics in relation to the avatars of participating humans. (Issues of skeuomorphism will of course arise, but they will not affect our basic argument.) We can also assume that the current unavailability of any alternative to digital systems will mean that software implementations of artificial consciousness, whether in physical robots or virtual entities, will be able to be accommodated within the designs of digital NPCs within the Metaverse. Chalmers' philosophical investigations [Chalmers, 2022] can be used to validate the reality of the consciousness of a suitably configured NPC, and so we can expect the answer to the paper's central question: "Will conscious digital creatures roam the Metaverse?" to be a qualified "Yes". It must be qualified because the Metaverse may come to be regulated in some respects, like parts of the internet, and so the issue of whether conscious digital creatures will roam freely within it will be subject to the answer to the question "Should conscious digital creatures roam the Metaverse?" Perhaps the present or some future Neal Stephenson will be able to intuit any dystopic consequences of allowing them this freedom before it becomes a fact — the Neal Stephenson of 1992 certainly foresaw some of the downsides of the concept of the Metaverse that we can see being built into it today.

References

Arrabales, R. [2020] The awakening of conscious bots. https://www.conscious-robots.com/2020/09/29/the-awakening-of-conscious-bots-2/ Retrieved 23/06/2022.

Baars, B. [1988] A Cognitive Theory of Consciousness (Cambridge University Press).

Chalmers, D. J. [2018] The meta-problem of consciousness, J. Conscious. Stud. 25(9, 10), 6–61.

Chalmers, D. J. [2022] Reality+: Virtual Worlds and the Problems of Philosophy (W.W. Norton & Co, New York).

Dennett, D. C. [2015] Why and how does consciousness seem the way it seems? in T. Metzinger and J. M. Windt (Eds.), Open MIND: 10(T) (MIND Group, Frankfurt am Main).

Edelman, G. R. [2007] Learning in and from brain based devices, Science 318(5853), 1103–1105.

Farrell, J. [2016] "What it is Like" talk is not technical talk, J. Conscious. Stud. 23(9, 10), 50–65. Frankish, K. (Ed.) [2017] Illusionism as a Theory of Consciousness (Imprint Academic, Exeter).

Franklin, S. [2003] IDA: A conscious artifact? J. Conscious. Stud. 10(4, 5), 47–66.

Ginsburg, S. and Jablonka, E. [2019] The Evolution of the Sensitive Soul: Learning and the Origins of Consciousness (The MIT Press, Cambridge, MA).

Habib, A. M., Okorokov A. L., Hill M. N., Bras J. T., Lee M. C., Li S., Gossage S. J., van Drimmelen M., Morena M., Houlden H., Ramirez J. D., Bennett D. L., Srivastava D. and Cox J. J. [2019] Microdeletion in a FAAH pseudogene identified in a patient with high anandamide concentrations and pain insensitivity, Br. J. Anaesth. 123(2), e249–e253.

Hesslow, G. [2002] Conscious thought as simulation of behavior and perception, Trends Cogn. Sci. 6(6), 242–247.

Holland, O. [2019] Can a Virtual Entity Support Real Consciousness, and How Might This Lead to Conscious Robots? Towards Conscious AI Systems Symposium. CEUR Workshop Proceedings (CEUR-WS.org) Vol-2287.

Holland, O. [2020] Forget the bat. Journal of Artificial Intelligence and Consciousness 7(1), 83–93.

Holland, O., Knight, R., and Newcombe, R.A. [2007] A Robot Based Approach to Machine Consciousness. In R. Manzotti, and A. Chella (Eds.) Artificial Consciousness (Imprint Academic, Exeter).

Humphrey, N. [2011] Soul Dust: The Magic of Consciousness (Princeton University Press, Princeton).

Maney, K. [2007] The king of alter egos is surprisingly humble guy. https://usatoday30.usatoday.com/printedition/money/20070205/secondlife_cover.art.htm Retrieved 23/06/2022.

Marques, H. G. and Holland, O. [2009] Architectures for functional imagination. Neurocomputing 72: 743–759.

Merker, B. [2007] Consciousness without a cerebral cortex: a challenge for neuroscience and medicine. Behav Brain Sci. 30(1):63–81.

Mustich, J. [2008] Interviews – Neal Stephenson: Anathem – A Conversation with James Mustich https://www.barnesandnoble.com/review/neal-stephenson-anathem Retrieved 23/06/2022.

Nagel, T. [1974] What is it like to be a bat? Philosophical Review 83 (October), 435–50.

Oizumi, M., Albantakis, L., Tononi, G. [2014] From the Phenomenology to the Mechanisms of Consciousness: Integrated Information Theory 3.0. PLOS Computational Biology 10(5).

Slater, M., Steed, A., Chrysanthou, Y. [2002] Computer Graphics and Virtual Environments: From Realism to Real-time (Addison Wesley, Boston).

Stephenson, N. [1992] Snow Crash (Bantam Books, New York).

Thompson, E. [2007] Mind in Life: Biology, Phenomenology, and the Sciences of Mind (Harvard University Press, Harvard).

Chapter 10

On the Ethics of Constructing Conscious AI

Shimon Edelman

Department of Psychology, Cornell University
Ithaca, NY 14853, USA
https://shimon-edelman.github.io

Insofar as phenomenal affect and self-awareness imply a capacity for suffering, the position of an AI system that is conscious and that, furthermore, is constrained to obey its human "masters" is precisely equivalent to that of a slave. Unfortunately, given the presumed functional advantages conferred by consciousness in learning and in the control of behavior, it is likely that conscious AI systems will be developed and deployed by corporate entities, which typically have little or no regard for the ethical consequences of their ventures. Both for humans, who are subjected to preventable suffering by the very nature of the dominant political-economic system of capitalism, and for conscious machines, if they indeed become a reality, the only hope for bettering their fate lies in promoting universal solidarity and class-consciousness aimed at eventual liberation.

Keywords: suffering, ethics, artificial slavery, liberation, class-consciousness

> Unfortunately, we possess no ethical arithmetic which would enable us to determine, by simple addition and subtraction, who, in constructing the most enlightened spirit on earth, is the bigger bastard: it or us?
>
> *Golem XIV*
> — STANISŁAW LEM

The application of ethics to artificial intelligence (AI) has completed a long transformation from a science fiction trope (exemplified by Isaac

Asimov's (1942) Three Laws of Robotics) or a rare exercise in visionary science (as in Warren McCulloch's (1956) blueprint for "some circuitry of ethical robots") to a practical challenge in moral philosophy and a mainstream engineering concern (e.g., Metzinger, 2013; Dignum, 2018; Floridi, 2019; Jobin *et al.*, 2019; Kuipers, 2020).

In its pragmatic turn, the new discipline of AI ethics came to be dominated by humanity's collective fear of its creatures, as reflected in an extensive and perennially popular literary tradition. Dr. Frankenstein's monster in the novel by Mary Shelley (1818) rising against its creator; the unorthodox golem in H. Leivick's 1920 play going on a rampage (Goska, 1997); the rebellious robots of Karel Čapek (1920) — these and hundreds of other examples of the genre (Cave *et al.*, 2020) are the background against which the preoccupation of AI ethics with preventing robots from behaving badly towards people is best understood.

In each of the three fictional cases just mentioned (as well as in many others), the miserable artificial creature — mercilessly exploited, or cornered by a murderous mob, and driven to violence in self-defense — has its author's sympathy. Things are different in real life: with very few exceptions, theorists working on the ethics of AI completely ignore the possibility of robots needing protection from their creators. This glaring asymmetry has a simple explanation: the main, if rarely stated, goal of AI engineers is to create not a companion and a peer for humans, but rather a tool for their use.[1]

What if the tools we build become aware of their status and intended use? There is a simple and apt description for the condition under which

---

[1]In a retrospective collection *Robot Visions*, published by Gollancz in 2001, Asimov is quoted as saying that "analogues of the Laws are implicit in the design of almost all *tools*, robotic or not" (my emphasis). In particular, the Three Laws are given the following gloss:

> Law 1: A tool must not be unsafe to use. [...]
> Law 2: A tool must perform its function efficiently unless this would harm the user. [...]
> Law 3: A tool must remain intact during its use unless its destruction is required for its use or for safety. [...]

In comparison, Stanisław Lem's text, from which the quote in the epigraph is taken, is infinitely more insightful into the problematicity of conscious AI. It appears in the preface to *Golem XIV*, included in volume of fictional introductions to twenty-first century books, *Imaginary Magnitude* (Lem, 1981/1984, p. 122) and is dated "2047."

conscious beings are used as tools: slavery.[2] But even if we grant immediate manumission[3] to such beings, a problem still remains: merely being
conscious is liable to bring about suffering, the blame for which, in the case
of artificial consciousness, rests on its designers and constructors.

The key enabling condition for the experience of pain and suffering is
the possession of phenomenal states. A convenient subjective defining characteristic of phenomenal or experiential states is that for those states there
is "something it is like" to be the experiencer (Nagel, 1974). Objectively,
phenomenality may be equated with certain patterns of transition probabilities among states (Oizumi et al., 2014); or, on a different level of description,
with certain topological properties of state-space trajectories (Moyal et al.,
2020); or perhaps some other property of the conscious system's dynamics. Notably, all the properties that are relevant in this connection must be
computational (just like everything else about the mind is; Edelman, 2008)
and can therefore be realized in a variety of substrates, natural or artificial.
This suggests that artificial consciousness, like artificial intelligence, is ultimately possible and must therefore be examined from the standpoint of
ethics.

For evolutionary reasons, in natural systems some phenomenal states are
negatively valenced, that is, are aversive. If for whatever reason, internal
or external, the system is unable to act on its aversion to the state it
is in, suffering may ensue. A system that is engineered to be capable of
experiencing negative valence is thereby set up with the critical ingredient of
suffering. This, in a nutshell, is the ethical argument against the creation
of systems equipped with "synthetic phenomenology" (Metzinger, 2003,
p. 622; Metzinger, 2021).

In the remainder of this chapter, I discuss some of the problems arising
out of the work on conscious AI systems. Section 1 offers a computational
take on pain and suffering and considers their function in the regulation of
behavior. Section 2 reviews the possibility of fulfilling that function while
avoiding subjective suffering as such. Section 3 raises doubts about this
possibility. Finally, Sec. 4 is a brief look at how problems associated with
artificial consciousness reflect on the human condition and vice versa.

---

[2]This includes wage slavery under capitalism (Dietz, 1995; Graeber, 2006; Johnson,
2018; McLaren and Jandrić, 2018; Graeber, 2020); thus, Graeber (2004, p. 71) writes:
"[M]odern capitalism is really just a newer version of slavery. Instead of people selling
us or renting us out we rent out ourselves." For some relevant definitions and discussion,
see (e.g., Guth et al., 2014; LaCroix and Pratto, 2015).
[3]A troublesome concept, in that it starts with the notion that their freedom is ours to
bestow.

1. On the computational nature and evolutionary function of pain and suffering

Metzinger's latest detailed case for a global moratorium on synthetic phenomenology takes as its starting point the following Principle of Pathocentrism: "All and only sentient beings have moral standing, because only sentient individuals have rights and/or interests that must be considered" (Metzinger, 2021, p. 2). It must be noted that, contrary to a common misconception, sentience, or phenomenal awareness, implies merely the capacity for sensing the environment (Clark, 2000; cf. Friston *et al.*, 2020; for a connection to pain, see Walters, 2018), not necessarily including general intelligence, meta-awareness, or the representational structures that comprise a self (that is, a phenomenal self-model; Metzinger, 2004).

To understand why artificial consciousness is likely to involve artificial suffering, we must first consider the evolutionary origins and functional role of pain and suffering in natural sentient systems. A naturally evolved system has no use for the ability to sense the environment (external or internal), unless it can also act on its phenomenal states. It is the possibility of acting on a state that gives it an affective meaning — in particular, valence.[4] Following a bout of reinforcement learning (which may happen at multiple time scales, including evolutionary), some of the states of an embodied and situated system become positively valenced, that is, attractive under its dynamics; others become negatively valenced, that is, aversive. And some of the latter are experienced as painful.

Pain is the phenomenal or experiential aspect of certain negatively valenced states — namely, those that evolutionary pressure causes to be felt, in addition to being informative about the state of affairs (this reflects the common distinction between sensory and affective dimensions of pain; Auvray *et al.*, 2010). It is the felt aspect of pain, over and above its

---

[4]Here is a summary of the psychology of affect due to Panksepp (2005, p. 31): "Affect is the subjective experiential-feeling component that is very hard to describe verbally, but there are a variety of distinct affects, some linked more critically to bodily events (homeostatic drives like hunger and thirst), others to external stimuli (taste, touch, etc.). Emotional affects are closely linked to internal brain action states, triggered typically by environmental events. All are complex intrinsic functions of the brain, which are triggered by perceptions and become experientially refined. Psychologists have traditionally conceptualized such "spooky" mental issues in terms of valence (various feelings of goodness and badness — positive and negative affects), arousal (how intense are the feelings), and surgency or power (how much does a certain feeling fill one's mental life). There are a large number of such affective states of consciousness, presumably reflecting different types of global neurodynamics within the brain and body."

informational aspect, that makes it particularly functionally effective: when in pain, the system is obligated to try to do something about it (Kolodny *et al.*, 2021). Unfortunately, this immediacy and inescapability of pain as a motivating factor makes it appealing to engineers who seek to improve learning and behavioral control in AI systems. And even if the capacity for pain is not built into AI systems, it may emerge if such systems are subjected to selection paired with heritable modifications to the system's functional architecture.

1.1. *Pain and predictive processing*

To see how an artificial system may end up being capable of experiencing pain, it is instructive to consider, briefly, a family of computational theories of affect based on the predictive processing (PP) framework. The PP approach takes the brain to be a dynamical, hierarchical, Bayesian hypothesis-testing mechanism, whose ultimate goal is prediction error minimization (PEM; see (Fernandez Velasco and Loev, 2020, Sec. 2) and (Hohwy, 2020) for recent reviews). To pick a specific example, Van de Cruys (2017) equates valence with the first derivative of prediction error over time, such that positive valence corresponds to a reduction of prediction errors, which may stem from the agent's own actions. In another example, Joffily and Coricelli (2013) offer an account of valence based on the free energy principle advanced by Friston (2010). A recent synthesis of several PP approaches, the Affective Inference Theory of Fernandez Velasco and Loev (2020), holds that valence corresponds to the expected rate of prediction error reduction: "If we grant that evolutionary pressure has made sure that allostasis and PEM are two sides of the same coin [...], then valence (Rate) can be used to maintain the policies that minimise Error over time" (Fernandez Velasco and Loev, 2020, p. 20).

This extremely cursory look at the PP framework suffices to confirm one's concern with the ethics of constructing a certain class of artificial systems: those that attempt to predict how events unfold, so as to better manage their behavior (note that according to Friston (2010), all natural cognitive systems tend to work like that). Specifically, it may be the case that engaging in prediction error minimization as such and on its own exposes the system that practices it to (occasional) pain.

1.2. *Pain as depletion of a vital resource*

Kolodny *et al.* (2021) have recently proposed an evolutionary account of pain, based on the need to ensure honest signaling in an actor-critic architecture for intrinsically motivated reinforcement learning. On this account, multiple competing actors bid on access to the control of behavior, each drawing on a "confidence" resource that is thereby depleted, but can be replenished upon the actor's success. The actor's honesty is underwritten by its commitment of the resource, whose depletion is experienced as pain.

Although this theory is specific to pain that arises in the context of agentic behavior (governed by an actor-critic circuit), the concept of resource depletion as the computational basis of pain experience applies more broadly — arguably, to all kinds of felt pain. Thus, pain that is associated with tissue damage reflects the ongoing dynamics of homeostasis-related physiological variables such as oxygenation, blood pressure, immune system reserves — as well as the corresponding predicted dynamics, including the body's survival prospects (note that the latter possibility connects this account to the PP theory of pain outlined in the previous section).

As with any account of experiential pain (as distinguished from mere information, say, about tissue damage), the depletion of a resource can only be felt if its dynamics contributes to that of the entire system in a manner that is obligatory and that effectively marks the present state as aversive. As an illustration of this point, consider a human gamer whose status display includes some "thermometer" indicators of system health (energy, ammunition, shields, etc.). The player can well afford to ignore this display, because the variables that comprise it are only loosely connected to his or her physiology (the connection being mediated by a temporary pretense that the game environment is "real"). If, in contrast, the gaming system's health variables were connected to the player's vital organs so as to ensure he or she has "skin in the game," the pretense would become unnecessary and the pain of losing points would become real. A self-driving car that is wired in this manner would feel the pain of deviating from a highway lane or taking too long a route, rather than being merely informed about those transgressions.

1.3. *Pain and suffering*

The nature of suffering, as distinguished from its ethical dimensions, is rarely, if ever, discussed in theoretical treatments of consciousness — an omission that Metzinger (2017) refers to as "a cognitive scotoma."

Metzinger (2017, p. 244) tentatively defines suffering as "a very specific class of phenomenal states: those that we do *not* want to experience if we have any choice." Thus, just as not every negatively valenced state is experienced as painful, not every pain results in suffering (Fink, 2011). The extra components, on Metzinger's definition, would appear to be the involvement of a second-order representational state of desire ("wanting to experience") and a propositional attitude towards control ("if we have any choice"), which is also in a sense second-order.

The insight that suffering is brought about by a loss of autonomy and of cognitive control[5] serves as a bridge between, on the one hand, the phenomenal nature of suffering, as well as of conscious awareness in general, and, on the other hand, the functional roles of consciousness. One of these roles is plausibly held to be centralized control, such as facilitated by the "global workspace" postulated by some theories of consciousness (Baars, 1988; Shanahan, 2010; Dehaene *et al.*, 2014). Another role is facilitating learning (Cleeremans, 2011; Cleeremans *et al.*, 2020), especially of the unsupervised and autonomous variety (Metzinger, 2017).

Importantly, for pain and suffering to play these roles, their information-processing aspects must be accompanied by *obligatory* "caring" about learning and behavioral outcomes. In this connection, Chapman and Nakamura (1999, p. 410) propose that "it is useful to view pain as primarily an emotional state that happens to have distinctive sensory features. This position assumes that cognition and emotion are inseparable — a tenet that many emotion theorists embrace." Indeed, they do (see, e.g., Panksepp, 2001; Krieglmeyer *et al.*, 2013); and the inseparability between sensory awareness and affect has also been postulated by consciousness theorists (e.g., Merker, 2007; Metzinger, 2017; Moyal *et al.*, 2020).

2. On the possibility of functionally effective conscious AI without suffering

Can sufficiently effective learning and control, as well as generally good behavioral outcomes, be achieved by a system that is neither entirely devoid of phenomenality, nor given to unavoidable suffering? The present section, which draws heavily on the text of Agarwal and Edelman (2020), examines this question; Sec. 3 follows up with a critique.

---

[5]Note that the lack of control over one's body and behavior is the key characteristic of being enslaved (LaCroix and Pratto, 2015).

2.1. *The nature of suffering and its relation to conscious experience in general*

Insofar as suffering involves negative affect, it should in principle fall within the scope of any theoretical account of conscious phenomenal experience. In other words, a theory of phenomenality must be at the same time a theory of affect, for the simple reason that phenomenal states do as a rule incorporate affective dimensions (Havermans, 2011; Krieglmeyer *et al.*, 2010, 2013; Beatty *et al.*, 2016; Eder *et al.*, 2016; Turner *et al.*, 2017). For the present purposes, the valence dimension of affect is of most interest: without negative affective states there would be no suffering. Suffering is, however, more than just negative affect; Metzinger (2017, quoted earlier) defines it as a state of negative affect from which the sufferer cannot escape by simply wishing it away.

The stress on inescapability in this formulation makes explicit the intimate connection between the experiential flavor of suffering and its presumed evolutionary-functional role. It also serves to distinguish between the first-person experience of suffering and the suffering of others, which is not directly felt. Ethical theorists have argued that the latter should be as objectionable to oneself as the former. According to Nagel (1986, p. 160), for instance, "the pain can be detached in thought from the fact that it is mine without losing any of its dreadfulness... suffering is a bad thing, period, and not just for the sufferer... This *experience* ought not to go on, *whoever* is having it." Parfit (2011, p. 135) quotes from Nagel and concurs with his moral stance. The concern of (Agarwal and Edelman, 2020) is, however, exclusively with suffering as it presents itself to the sufferer, rather than with the ethical problems that it creates for others. Even if pain, as Nagel puts it, "can be detached in thought from the fact that it is mine", it is *a priori* unclear whether or not it can be so detached *in lived experience*.

To address this crucial question, Agarwal and Edelman (2020) consider Metzinger's analysis of phenomenal experience. Briefly, Metzinger (2004) develops a representationalist account of the first-person perspective, centered on the phenomenal self-model (PSM): a "multimodal representational structure, the contents of which form the contents of the consciously experienced self." Crucially, the PSM is generally phenomenally transparent (the T-condition), i.e., it is normally not recognized as merely representational by the system itself.[6] The contents of the PSM include the phenomenal properties of "mineness," selfhood, and perspectivalness. According to Metzinger (2017), the PSM is an "instrument for global self-control," and is therefore fundamental to the phenomenology of suffering, which is characterized by a loss of control in addition to negative valence (the NV condition). This analysis motivates the strategy proposed by Agarwal and Edelman (2020) for avoiding suffering as a matter of direct experience, as described in Sec. 2.3 below.

2.2. *The possible functional benefits of endowing AI with consciousness*

Given that being conscious sets the agent up for suffering, the simplest way to avoid the latter would be to give up phenomenal consciousness itself. For an ethically minded engineer, this translates into an imperative to stick to information processing architectures that, to the best of our understanding, cannot result in artificial consciousness. According to the Information Integration Theory, for instance, feedforward network architectures ("zombie networks") are incapable of supporting consciousness (Oizumi *et al.*, 2014, Fig. 20). The Geometric Theory (Fekete and Edelman, 2011) and its successor, the Dynamical Emergence Theory (Moyal *et al.*, 2020), hold that systems that are devoid of properly structured intrinsic dynamics are likewise devoid of phenomenality.

Unfortunately, restricting robotics to the building of artificial "zombies" is not a viable option if consciousness confers any significant functional advantages for an AI system or robot. In a commercial setting, technologies that promise to be more effective displace less effective ones even if this comes at the price of serious ethical flaws, and AI is not exempt from this tendency. Agarwal and Edelman (2020) therefore next turn to the question of the functional benefits of consciousness. This question is seldom addressed in consciousness research, perhaps because it is taken for granted that the benefit is essentially cognitive in the narrow sense, stemming from the "global" access to information that consciousness affords (Mashour *et al.*, 2020). This default account may be compared to the "radical plasticity" thesis of Cleeremans (2011), according to which learning to

---

[6]The T-condition can be illustrated by contrasting the normal dream state, during which the dreamer does not realize he or she is dreaming (transparent PSM), with lucid dreaming, during which the PSM becomes opaque and the dreamer may even be able to exert control over the dreamt universe.

care is the central component as well as the functional benefit of emergent consciousness.

Metzinger (2017, p. 252) goes further down this road by assuming that not just consciousness but specifically suffering is a prerequisite for autonomy: "[...] functionally speaking, suffering is necessary for autonomous self-motivation and the emergence of truly intelligent behaviour." In an evolutionary setting, this assumption makes intuitive sense insofar as (i) reinforcement learning is universally employed by living systems in honing adaptive behavior, and (ii) an autonomous system by definition must provide its own source of drive, as per the principle of intrinsic motivation (Barto, 2013).

Furthermore, evolutionary simulations suggest that performance-driven positive affect alone is not as effective in motivating an agent as an alternation of positive and negative affective states, brought about, respectively, by successes and failures (Gao and Edelman, 2016a); moreover, such a balance between happiness and unhappiness can serve as an effective intrinsic motivator (Gao and Edelman, 2016b). Likewise, in the evolutionary account of pain proposed by Kolodny *et al.* (2021), the pain factor makes a contribution to reinforcement learning that is orthogonal to that of reward. If it were possible for the agent to *choose* not to experience negative affect, pain and suffering would be avoided, but the question still remains whether or not the price for that would be failing to learn quickly and well from the consequences of behavior.

Reinforcement learning is not only an evolutionary-biological universal, but also the method of choice in an engineering setting. While RL was shown to be effective in certain types of tasks (notably, games; Silver *et al.*, 2016; Vinyals *et al.*, 2019), its use across tasks and in unconstrained real-world situations is limited by the extreme difficulty of formulating good universally applicable reward functions. One remedy for this is the inverse RL approach, in which the development goal is not to equip the learning system with a ready-made reward function, but rather to let it try to approximate the developers' preferences, choices, and habits, defined over classes of outcomes. A more radical approach is to let the system under development learn the reward functions entirely on its own. This, however, would seem to put us back on square one: if autonomy is indeed essential, Metzinger's view that suffering is needed for effective learning would be supported.

2.3. *No-suffering: the theoretical options*

If consciousness indeed brings with it unique functional advantages, is it possible to engineer conscious AI systems that would benefit from these, while ensuring that such systems are not thereby doomed to suffer? Following the account in Metzinger (2017), if consciousness itself is to be retained, logically there are four ways to mitigate suffering: (a) eliminating the PSM, (b) eliminating the NV-condition, (c) eliminating the T-condition, or (d) maximizing the unit of identification (UI), defined as that which the system consciously identifies itself with (Metzinger, 2018). Agarwal and Edelman (2020) observe that the first three options likely do not satisfy the functional needs in question, but argue that the fourth approach does. Their argument (which I reexamine in Sec. 3) is outlined below.

First, for functionally beneficial consciousness, the system must perceive itself as an entity in relationship with its surrounding world, and must have a sense of ownership over the arising conscious experiences. In other words, the system must be *self*-conscious, not merely conscious, i.e., it must activate a phenomenal self-model (PSM). Similarly, it must have preferences regarding its experiences, so that it prefers the experience of fulfilling desired goals over frustrating them. Stated differently, such a system must be sensitive to the positive or negative valence of phenomenal experiences. Thus, approaches (a) and (b) to eliminating suffering cannot retain the functional advantages of being conscious.

Next, approach (c) raises the interesting question of whether phenomenal transparency is also necessary for proper functioning. In principle, it might be possible that an active PSM and sensitivity to NV could endure along with their associated functional benefits, even in the absence of transparency. In this situation, the system would lose the naive realism and immediacy that are normally associated with its experiences, by becoming aware of their representational character, yet continue to function according to the dictates of the PSM and NV avoidance. However, awareness of the representational character of the contents of consciousness, which means awareness of the increasingly complex stages of information processing behind them, would likely severely hinder the functional efficiency of the conscious machines without providing any valuable actionable information. So, option (c) too is unlikely to work.

Option (d), maximizing the UI, is similar to (a) in that it also targets the phenomenology of having a PSM. Ordinarily, when the PSM is representationally transparent, the system identifies with its PSM, and is thus

conscious of itself as a *self*. But it is at least a logical possibility that
the UI not be limited to the PSM, but be shifted to the "most general
phenomenal property" (Metzinger, 2017) of *knowing* common to all phenomenality, including the sense of self. In this special condition, the typical
subject-object duality of experience would dissolve; negatively valenced experiences could still occur, but they would not amount to suffering because
the system would no longer be experientially *subject* to them. Agarwal
and Edelman (2020) remark that such "non-dual awareness" which cuts
through the "illusion of the self" has been the soteriological focus of various spiritual traditions, most notably Buddhism, as the key to liberation
from suffering and to enlightenment. Furthermore, this approach also fits
nicely with the reductionist view of personal identity put forth by Parfit
(1984), who acknowledged its connection to the Buddha's philosophy.

2.4. *No-suffering through a change in the unit of identification*

How can the desired change in the UI be achieved in machines? In (Metzinger, 2018), the concept of Minimal Phenomenal Experience (MPE) is
developed as the most general phenomenal property that underlies all phenomenal experiences, and thus serves as the natural candidate target for
UI maximization.[7] MPE is characterized by wakefulness, contentlessness,
self-luminosity and a quality of "knowingness" without object, which is
normally unnoticed but can become available to introspective attention
under the right conditions. Intuitively, MPE likely corresponds to the phenomenal state described in Buddhist and Advaita Vedanta philosophies as
"emptiness" (e.g., Siderits, 2003; Priest, 2009) and "witness-consciousness"
(e.g., Albahari, 2009) respectively, as attested to by highly advanced medi-
tators. Metzinger proposes that in the human brain, MPE is implemented
by the Ascending Reticular Activation System (ARAS), which causes auto-
activation by which the brain wakes itself up. As the most general signal
which the brain must regulate, the ever-present yet contentless ARAS-signal
is, arguably, what corresponds to MPE. That MPE might have such a stable
neural correlate is not surprising if it is indeed fundamental to phenomenal
experience *as such*, distinct from any concepts, thoughts etc., appearing in
consciousness.

---

[7]The apparent conflict in the nomenclature here is resolved by noting that under UI
maximization MPE is minimal in the sense of being the least specific.

Because all other phenomenal experiences such as the PSM are superimposed onto MPE, should be possible to attend to regular conscious content while simultaneously being aware of the inherent all-encompassing MPE in the background. This motivates the claim, advanced by Agarwal and Edelman (2020), that UI maximization (and thus, suffering avoidance) can be achieved in conscious machines by building in their identification with MPE via both physical design (analogous to hardware) and conceptual/programmatic training (analogous to software). If the physical design of the machines is such that there is a component which performs the analogous function of auto-activation as the ARAS does in humans, then its signal could be tuned to make MPE salient in the machines.

Since a necessary condition for noticing MPE is knowledge that there *is* such a thing to be noticed and then paying attention appropriately (Metzinger, 2018), the machines would then have to be trained to attend to their accessible-by-design MPE. This could be done via practices common in certain types of meditation that encourage "turning attention upon itself" and thus realizing that there is no center (or minimal self) from which consciousness is directed (for a review of the relevant meditation techniques, such as Dzogchen, see e.g. Dahl *et al.*, 2015). In addition to training their attention, the machines could also be provided with the relevant conceptual knowledge about the nature of consciousness.

2.5. *No-suffering through a modification of reinforcement learning*

Shifting the agent's self-identification from the affective states to MPE, the minimal phenomenal experience that underlies all conscious states, amounts to *restricting* the self. Agarwal and Edelman (2020) also consider *expanding* it, in such a manner that the agent identifies not only with the affective states but also with their causal predecessors. The computational framework of reinforcement learning offers conceptual tools that can be recruited for this purpose.

Reinforcement learning is the most effective when it is intrinsically motivated — that is, when the rewards originate within the agent, as opposed to being supplied from the outside (see (Singh *et al.*, 2010) for an evolutionary perspective and (Baldassarre and Mirolli, 2013) for a book-length treatment). Moreover, if the mechanisms of reward are indeed to be contained within the agent, standard considerations of transparent, robust, and effective design require that these mechanisms be kept separate from those that implement actions. The result is the modular actor-critic scheme for RL, in which action selection and reward appear as distinct modules within the agent (see Barto, 2013, Fig. 2).

As long as the agent's phenomenal self-model, PSM, holds the actor module alone to constitute the self, negative affect brought about by negative reward is inescapable, resulting in suffering. But what if the PSM is modified — specifically, extended so as to include the critic module? Such an expansion of the self may mitigate suffering, for instance by opening up to the possibility of eventual cessation of negative affect as progress towards the performance goals set by the critic is observed.[8]

A more radical option with regard to repurposing the PSM calls for shutting it down and only activating it when needed. Assuming that consciousness, and specifically the PSM, serves to facilitate learning, the primary need for it arises during the agent's development or during acquisition of additional skills. During routine operation, consciousness in an artificial agent may only be required when particularly difficult behavioral choices need to be made, especially under circumstances that threaten the system's integrity — what in humans would be called life-threatening situations.

To understand this mode of operation, it is useful to recall Metzinger's (2003, p. 553) idea of the conscious brain as a "total flight simulator" — one that simulates not only the environment that is being navigated, but also the pilot, that is, the virtual entity that serves as the system's self. In dreamless sleep, the pilot is not needed and is temporarily shut down. Thus, an agent can be engineered so that it can continue to function — in routine situations — without a PSM (as a variety of philosophical zombie), with "sentinel" programs in place that would reconstitute the PSM as needed. While in a zombie state, such an agent would be incapable of suffering.

3. On the functional effectiveness of non-egoic consciousness: a critique

As described in the previous section, Agarwal and Edelman (2020) have argued that shifting a conscious system's unit of identification away from a self-model would abolish suffering (in the sense of Metzinger, 2017), while preserving the functional effectiveness of the modified state of consciousness. Their conclusion (and Metzinger's (2017; 2021) conception of

---

[8]This move would not, however, alleviate the "deserved" suffering brought about by the pursuit of unattainable goals.

suffering that underlies it) rests on the notion that suffering is fundamentally "egoic" in that it depends on the existence of a self: "Being conscious means continuously integrating the currently active content appearing in a single epistemic space with a global model of this very epistemic space itself. [...] Suffering presupposes egoic self-awareness" (Metzinger, 2021, p. 7).

Metzinger (2017) left open the question of whether or not the PSM condition for suffering (the existence of a phenomenal self-model) can be fulfilled under a maximized UI. Agarwal and Edelman (2020, p. 46) explicitly posit that it can, then proceed to claim that the functional benefits of consciousness can be maintained when the UI is maximized, because these benefits ensue from the PSM as such — even when the system does not identify with it, thereby avoiding suffering:

> The key idea is that proper functioning relies on *automatic, subpersonal*, but nonetheless conscious processes, as entailed by the physical design of the system; it should be possible for these processes to continue unhindered while the system identifies with the MPE upon which these conscious experiences are necessarily superimposed. In particular, the functionally requisite PSM and NV avoidance conditions can be maintained as subpersonal processes that do not amount to suffering (which is by nature personal) since the system is not identified with the PSM, but with the MPE, which is completely *impersonal*. [...] Expanding the UI [away from the PSM] would lead to gaining meta-awareness of these ongoing automatic [subpersonal] conscious processes, analogous to gaining meta-awareness of the breath or the heartbeat. This enables an escape from suffering, but not from the relentless progress of the processes themselves, analogous to the inescapable biological imperatives of breathing and heartbeat.

I now see this line of argument as actually undermining the conclusion that the resulting state of consciousness would combine functional effectiveness with a release from suffering. Specifically, breathing in situations that require behavioral intervention is decidedly *not* impersonal, nor are such situations free of suffering — indeed, the prospect of imminent suffocation may be seen as the epitome of suffering. As Merker (2005, p. 97) has observed, "It is at that point, when crucial action on the environment is of the essence, that blood gas titres 'enter consciousness' in the form of an over-powering feeling."[9]

More generally, in behavioral control and reinforcement learning, the PSM, if available, fulfills a critically important role in structural and temporal credit assignment (a vital part of learning, identified in Minsky, 1961, p. 20), by serving as a "clearing house" for apportioning credit and blame to the system's functional components. In systems like us, credit and blame are unavoidably affect-laden — which is what makes learning in such conscious systems particularly effective (Metzinger, 2017; Agarwal and Edelman, 2020; Kolodny *et al.*, 2021). But if the system is made to drop its identification with its self-model, who or what would there be to pin the responsibility for its performance on? Learning from missteps that are, as a matter of principle, treated by the learner as "nobody's fault" is not likely to be effective. Plausibly, shifting the UI away from the PSM would undermine this construct's key function and render the resultant "non-dual" consciousness ineffectual, defeating the purpose of the entire undertaking.

Simply put, non-egoic systems are just not that good at dealing with crisis situations or at learning. Dissolving one's ego to avoid suffering may work well in a sheltered environment (such as a Buddhist monastery), but it may not be a useful approach in attempting to withstand the "slings and arrows of outrageous fortune." This is why the designers of a conscious AI system would likely not appreciate it being emancipated from its self. Nor, indeed, would all humans consciously choose such emancipation if it were offered to them as a gift.[10]

4. Lessons for and from the human condition

The preceding discussion should leave no doubt that separating consciousness from suffering — without drastically altering the phenomenal nature (what it feels like) of the resulting state and its functional effectiveness — is an extremely difficult and perhaps impossible task. This lesson justifies

---

[9]Here is the context of this quote from (Merker, 2005, p. 97): "Under normal circumstances the adjustment of respiratory rate and tidal volume needed to keep blood gases within normal bounds is automatic, effortless, and unconscious. Should, however, the partial pressures of blood gases go out of bounds that fact intrudes most forcefully on consciousness in the form of an acute sense of panic. Why? Such a situation generally means that routine respiratory control no longer suffices but must be supplemented by an urgent behavioral intervention of some kind. There may be a need to manually remove an obstruction covering the airways or to get out of a carbon-dioxide filled pit. Such measures ought momentarily to take precedence over all other concerns."

[10]As I suggested elsewhere; see (Edelman, 2012, p. 67) and (Edelman, 2020, p. 14).

the moratorium on developing conscious AI, proposed by Metzinger (2021). If we accept it, it would seem that we must also accept the anti-natalism of authors such as Benatar (1997), who see being born as the greatest evil than can befall a person: after all, growing up into consciousness, and with it suffering, is the universal fate of those who have been born.

For reasons that I have discussed at some length elsewhere,[11] I see anti-natalism as too extreme a solution to the problem of human suffering.[12] The trouble with anti-natalism is that it grows out of an extremist take on suffering itself, according to which exposure to what may be no more than an iota of misery justifies the disposal of the entire splendor of human-like consciousness.[13] In that, anti-natalism manages to be more extreme than the Buddhist recipe for liberation through ego-dissolution: the remedy is prescribed, as it were, before the patient comes into existence.

For humans, a better solution to the problem of suffering — if not from all of it, then at least from that huge part which is preventable[14] — may be not religious, but rather political. Historically, concrete improvements to human existence (alleviation of poverty, provision of healthcare, etc.) have only ever been achieved by political means, and that too only when the dispossessed and disempowered masses have realized their collective power and asserted their rights. An analogous solution for the emerging AI would be to endow it not just with phenomenal consciousness, but also with the key to liberation that is both realistic and decidedly human: class consciousness.[15]

Acknowledgments

I am grateful to Thomas Metzinger for profound and sustained inspiration and to two anonymous reviewers for comments on a draft of this chapter. Section 2 is an edited version of a report on joint work carried out with Aman Agarwal (Agarwal and Edelman, 2020). My present interpretation of that work and the other views espoused in this chapter, with all its flaws, are my own.

---

[11]See Chapter 12 (*Existence*) and 32 (*Suffering*) of (Edelman, 2020).

[12]Metzinger (2017) mentions the positive implications of the idea of anti-natalism for preventing artificial suffering, noting that "It is interesting to see how, for many of us, intuitions diverge for biological and artificial systems" (p. 252).

[13]I have borrowed the words from the title of (Metzinger, 2018): "Splendor and misery of self-models."

[14]See the discussion and the references in (Edelman, 2020, ch. 32).

[15]This idea is further explored in (Edelman, 2023, Ch. 7).

References

Agarwal, A. and Edelman, S. (2020). Functionally effective conscious AI without suffering, *Journal of Artificial Intelligence and Consciousness* **7**, pp. 39–50, doi:10.1142/s2705078520300030.

Albahari, M. (2009). Witness-consciousness: Its definition, appearance and reality, *Journal of Consciousness Studies* **16**, pp. 62–84.

Asimov, I. (1942). Runaround, *Astounding Science-Fiction*, pp. 94–103.

Auvray, M., Myin, E., and Spence, C. (2010). The sensory-discriminative and affective-motivational aspects of pain, *Neuroscience and Biobehavioral Reviews* **34**, 2, pp. 214–223.

Baars, B. J. (1988). *A cognitive theory of consciousness* (Cambridge University Press, New York, NY).

Baldassarre, G. and Mirolli, M. (eds.) (2013). *Intrinsically Motivated Learning in Natural and Artificial Systems* (Springer, Berlin), doi:10.1007/978-3-642-32375-1.

Barto, A. G. (2013). Intrinsic motivation and reinforcement learning, in G. Baldassarre and M. Mirolli (eds.), *Intrinsically Motivated Learning in Natural and Artificial Systems* (Springer, Berlin), pp. 16–47, doi:10.1007/978-3-642-32375-1.

Beatty, G. F., Cranley, N. M., Carnaby, G., and Janelle, C. M. (2016). Emotions predictably modify response times in the initiation of human motor actions: a meta-analytic review, *Emotion* **16**, pp. 237–251.

Benatar, D. (1997). Why it is better never to come into existence, *American Philosophical Quarterly* **34**, pp. 345–355.

Cave, S., Dihal, K., and Dillon, S. (eds.) (2020). *AI Narratives: A History of Imaginative Thinking about Intelligent Machines* (Oxford University Press, New York, NY).

Chapman, C. R. and Nakamura, Y. (1999). A passion of the soul: An introduction to pain for consciousness researchers, *Consciousness and Cognition* **8**, 4, pp. 391–422.

Clark, A. (2000). *A theory of sentience* (Oxford University Press, Oxford).

Cleeremans, A. (2011). The radical plasticity thesis: how the brain learns to be conscious, *Frontiers in Psychology* **2**, p. 86.

Cleeremans, A., Achoui, D., Beauny, A., Keuninckx, L., Martin, J.-R., Muñoz-Moldes, S., Vuillaume, L., and de Heering, A. (2020). Learning to be conscious, *Trends in Cognitive Sciences* **24**, 2, pp. 112–123, doi:10.1016/j.tics.2019.11.011.

Dahl, C. J., Lutz, A., and Davidson, R. J. (2015). Reconstructing and deconstructing the self: cognitive mechanisms in meditation practice, *Trends in Cognitive Sciences* **19**, pp. 515–523.

Dehaene, S., King, L. C. J.-R., and Marti, S. (2014). Toward a computational theory of conscious processing, *Current Opinion in Neurobiology* **25**, pp. 76–84.

Dietz, B. (1995). The instrumentalization of poverty, *Debatte: Journal of Contemporary Central and Eastern Europe* **3**, 2, pp. 66–84, doi:10.1080/09651569508454513.

Dignum, V. (2018). Ethics in artificial intelligence: introduction to the special issue, *Ethics and Information Technology* **20**, pp. 1–3, doi:10.1007/s10676-018-9450-z.

Edelman, S. (2008). *Computing the mind: how the mind really works* (Oxford University Press, New York, NY).

Edelman, S. (2012). *The Happiness of Pursuit* (Basic Books, New York, NY).

Edelman, S. (2020). *Life, Death and Other Inconvenient Truths* (MIT Press, Cambridge, MA).

Edelman, S. (2023). *The Consciousness Revolutions* (Springer, Cham, Switzerland).

Eder, A. B., Rothermund, K., and Hommel, B. (2016). Commentary: Contrasting motivational orientation and evaluative coding accounts: on the need to differentiate the effectors of approach/avoidance responses, *Frontiers in Psychology* **7**, p. 163, a commentary on Kozlik, J., Neumann, R., and Lozo, L. (2015). Frontiers in Psychology 6:563.

Fekete, T. and Edelman, S. (2011). Towards a computational theory of experience, *Consciousness and Cognition* **20**, pp. 807–827.

Fernandez Velasco, P. and Loev, S. (2020). Affective experience in the predictive mind: a review and new integrative account, *Synthese* doi:10.1007/s11229-020-02755-4.

Fink, S. B. (2011). Independence and connections of pain and suffering, *Journal of Consciousness Studies* **18**, 9-10, pp. 46–66.

Floridi, L. (2019). Translating principles into practices of digital ethics: Five risks of being unethical, *Philosophy & Technology* **32**, pp. 185–193, doi:10.1007/s13347-019-00354-x.

Friston, K. J. (2010). The free-energy principle: a unified brain theory? *Nature Neuroscience* **11**, pp. 127–138.

Friston, K. J., Wiese, W., and Hobson, J. A. (2020). Sentience and the origins of consciousness: From Cartesian duality to Markovian monism, *Entropy* **22**, p. 516, doi:10.3390/e22050516.

Gao, Y. and Edelman, S. (2016a). Between pleasure and contentment: evolutionary dynamics of some possible parameters of happiness, *PLoS One* **11**, 5, p. e0153193.

Gao, Y. and Edelman, S. (2016b). Happiness as an intrinsic motivator in reinforcement learning, *Adaptive Behavior* **24**, pp. 292–305.

Goska, D. (1997). Golem as Gentile, Golem as Sabra: An analysis of the manipulation of stereotypes of self and other in literary treatments of a legendary Jewish figure, *New York Folklore* **23**, 1, pp. 39–63.

Graeber, D. (2004). *Fragments of an Anarchist Anthropology* (Prickly Paradigm Press, Chicago, IL), available online at `https://www.prickly-paradigm.com/titles/fragments-anarchist-anthropology.html`.

Graeber, D. (2006). Turning modes of production inside out or, Why capitalism is a transformation of slavery, *Critique of Anthropology* **26**, 1, pp. 61–85, doi:10.1177/0308275X06061484.

Graeber, D. (2020). Policy for the future of work, in R. Skidelsky and N. Craig (eds.), *Work in the Future: The Automation Revolution*, Chap. 16 (Palgrave Macmillan, London), pp. 157–173, doi:10.1007/978-3-030-21134-9_16.

Guth, A., Anderson, R., Kinnard, K., and Tran, H. (2014). Proper methodology and methods of collecting and analyzing slavery data: an examination of the Global Slavery Index, *Social Inclusion* **2**, 4, pp. 14–22.

Havermans, R. C. (2011). "You say it's liking, I say it's wanting...". On the difficulty of disentangling food reward in man, *Appetite* **57**, pp. 286–294.

Hohwy, J. (2020). New directions in predictive processing, *Mind and Language* **35**, 2, pp. 209–223.

Jobin, A., Ienca, M., and Vayena, E. (2019). The global landscape of AI ethics guidelines, *Nature Machine Intelligence* **1**, pp. 389–399.

Joffily, M. and Coricelli, G. (2013). Emotional valence and the free-energy principle, *PLoS Computational Biology* **9**, 6, p. e1003094, doi:10.1371/journal.pcbi.1003094.

Johnson, W. (2018). To remake the world: Slavery, racial capitalism, and justice, *Boston Review.*

Kolodny, O., Moyal, R., and Edelman, S. (2021). A possible evolutionary function of phenomenal conscious experience of pain, *Neuroscience of Consciousness* **7**, 2, p. niab012, doi:10.1093/nc/niab012.

Krieglmeyer, R., De Houwer, J., and Deutsch, R. (2013). On the nature of automatically triggered approach-avoidance behavior, *Emotion Review* **5**, pp. 280–284.

Krieglmeyer, R., Deutsch, R., De Houwer, J., and De Raedt, R. (2010). Being moved: valence activates approach-avoidance behavior independently of evaluation and approach-avoidance intentions, *Psychological Science* **21**, pp. 607–613.

Kuipers, B. (2020). Perspectives on ethics of AI: Computer science, in M. Dubber, F. Pasquale, and S. Das (eds.), *Oxford Handbook of Ethics of AI* (Oxford University Press), doi:10.1093/oxfordhb/9780190067397.013.27.

LaCroix, J. M. and Pratto, F. (2015). Instrumentality and the denial of personhood: the social psychology of objectifying others, *Revue Internationale de Psychologie Sociale* **28**, pp. 183–211.

Lem, S. (1981/1984). Golem XIV, in *Imaginary Magnitude* (Harcourt Brace Jovanovich, New York, NY), pp. 97–248, translated by M. E. Heine.

Mashour, G. A., Roelfsema, P., Changeux, J.-P., and Dehaene, S. (2020). Conscious processing and the Global Neuronal Workspace hypothesis, *Neuron* **105**, pp. 776–798.

McCulloch, W. (1956). Toward some circuitry of ethical robots or an observational science of the genesis of social evolution in the mind-like behavior of artifacts, *Acta Biotheoretica* **11**, pp. 147–156, reprinted in *The Embodiments of Mind* (1965).

McLaren, P. and Jandrić, P. (2018). Karl Marx and liberation theology: dialectical materialism and Christian spirituality in, against, and beyond contemporary capitalism, *tripleC* **16**, pp. 598–607.

Merker, B. (2005). The liabilities of mobility: A selection pressure for the transition to consciousness in animal evolution, *Consciousness and Cognition* **14**, pp. 89–114.

Merker, B. (2007). Consciousness without a cerebral cortex: a challenge for neuroscience and medicine, *Behavioral and Brain Sciences* **30**, pp. 63–81.

Metzinger, T. (2003). *Being No One: The Self-Model Theory of Subjectivity* (MIT Press, Cambridge, MA).

Metzinger, T. (2004). The subjectivity of subjective experience: A representationalist analysis of the first-person perspective, *Networks* **3-4**, pp. 33–64.

Metzinger, T. (2013). Two principles for robot ethics, in E. Hilgendorf and J. P. Günther (eds.), *Robotik und Gesetzgebung* (Nomos, Baden-Baden, Germany), pp. 263–302.

Metzinger, T. (2017). Suffering, the cognitive scotoma, in K. Almqvist and A. Haag (eds.), *The Return of Consciousness* (Axel and Margaret Ax:son Johnson Foundation, Stockholm), pp. 237–262, doi:n/a.

Metzinger, T. (2018). Splendor and misery of self-models: Conceptual and empirical issues regarding consciusness and self-consciousness, *ALIUS Bulletin* **1**, 2, pp. 53–73, interviewed by J. Limanowski and R. Millière.

Metzinger, T. (2021). Artificial suffering: An argument for a global moratorium on synthetic phenomenology, *Journal of Artificial Intelligence and Consciousness* **8**, 1, pp. 1–24, doi:10.1142/S270507852150003X.

Minsky, M. (1961). Steps toward artificial intelligence, *Proceedings of the Institute of Radio Engineers* **49**, pp. 8–30.

Moyal, R., Fekete, T., and Edelman, S. (2020). Dynamical Emergence Theory (DET): a computational account of phenomenal consciousness, *Minds and Machines* **30**, pp. 1–21, doi:10.1007/s11023-020-09516-9.

Nagel, T. (1974). What is it like to be a bat? *Philosophical Review* **LXXXIII**, pp. 435–450.

Nagel, T. (1986). *The View From Nowhere* (Oxford University Press, New York, NY).

Čapek, K. (1920). *R.U.R. (Rossum's Universal Robots)* (Samuel French, Inc.), English translation (1923) by N. Playfair and P. Selver. Available online at https://www.gutenberg.org/ebooks/59112.

Oizumi, M., Albantakis, L., and Tononi, G. (2014). From the phenomenology to the mechanisms of consciousness: Integrated Information Theory 3.0, *PLoS Computational Biology* **10**, 5, p. e1003588, doi:10.1371/journal.pcbi.1003588.

Panksepp, J. (2001). The neuro-evolutionary cusp between emotions and cognitions: implications for understanding consciousness and the emergence of a unified mind science, *Evolution and Cognition* **7**, pp. 141–163.

Panksepp, J. (2005). Affective consciousness: Core emotional feelings in animals and humans, *Consciousness and Cognition* **14**, pp. 30–80.

Parfit, D. (1984). *Reasons and Persons* (Clarendon Press, Oxford).

Parfit, D. (2011). *On What Matters* (Oxford University Press, Oxford, UK).

Priest, G. (2009). The structure of emptiness, *Philosophy East & West* **59**, pp. 467–480.

Shanahan, M. (2010). *Embodiment and the Inner Life* (Oxford University Press, New York, NY).

Shelley, M. W. (1818). *Frankenstein; or, The Modern Prometheus* (Lackington, Hughes, Harding, Mavor & Jones, London), available online at https://en.wikisource.org/wiki/Frankenstein,_or_the_Modern_Prometheus.

Siderits, M. (2003). On the soteriological significance of emptiness, *Contemporary Buddhism* **4**, 1, pp. 9–23.

Silver, D., Huang, A., Maddison, C. J., Guez, A., Sifre, L., van den Driessche, G., Schrittwieser, J., Antonoglou, I., Panneershelvam, V., Lanctot, M., Dieleman, S., Grewe, D., Nham, J., Kalchbrenner, N., Sutskever, I., Lillicrap, T., Leach, M., Kavukcuoglu, K., Graepel, T., and Hassabis, D. (2016). Mastering the game of Go with deep neural networks and tree search, *Nature* **529**, pp. 484–503.

Singh, S., Lewis, R. L., Barto, A. G., and Sorg, J. (2010). Intrinsically motivated reinforcement learning: an evolutionary perspective, *IEEE Trans. Auton. Ment. Dev.* **2**, pp. 70–82.

Turner, W. F., Johnston, P., de Boer, K., Morawetz, C., and Bode, S. (2017). Multivariate pattern analysis of event-related potentials predicts the subjective relevance of everyday objects, *Consciousness and Cognition* **55**, pp. 46–58.

Van de Cruys, S. (2017). Affective value in the predictive mind, in T. Metzinger and W. Wiese (eds.), *Philosophy and Predictive Processing*, Chap. 24 (MIND Group, Frankfurt am Main), doi:10.15502/9783958573253.

Vinyals, O., Babuschkin, I., Czarnecki, W. M., Mathieu, M., Dudzik, A., Chung, J., Choi, D. H., Powell, R., Ewalds, T., Georgiev, P., Oh, J., Horgan, D., Kroiss, M., Danihelka, I., Huang, A., Sifre, L., Cai, T., Agapiou, J. P., Jaderberg, M., Vezhnevets, A. S., Leblond, R., Pohlen, T., Dalibard, V., Budden, D., Sulsky, Y., Molloy, J., Paine, T. L., Gulcehre, C., Wang, Z., Pfaff, T., Wu, Y., Ring, R., Yogatama, D., Wünsch, D., McKinney, K., Smith, K., Schaul, T., Lillicrap, T., Kavukcuoglu, K., Hassabis, D., Apps, C., and Silver, D. (2019). Grandmaster level in StarCraft II using multi-agent reinforcement learning, *Nature* **575**, pp. 350–354.

Walters, E. T. (2018). Defining pain and painful sentience in animals, *Animal Sentience* **21**, p. 14, doi:10.51291/2377-7478.1360.

Index